CHANCE AND CHAOS

DAVID RUELLE

Chance and Chaos

PRINCETON UNIVERSITY PRESS

Published by Princeton University Press, 41 William Street,
Princeton, New Jersey 08540
In the United Kingdom: Princeton University Press, Chichester, West Sussex

Library of Congress Cataloging-in-Publication Data

Ruelle, David.
Chance and chaos / David Ruelle.
p. cm.
Includes bibliographical references (p.).
ISBN 0-691-08574-9
ISBN 0-691-02100-7 (pbk.)
1. Probabilities. 2. Stochastic processes.
3. Chaotic behavior in systems. I. Title.
QA273.R885 1991
519.2—dc20 91-9564

First paperback printing, for the
Princeton Science Library, 1993

This book has been composed in Linotron Times Roman

Princeton University Press books are printed on
acid-free paper and meet the guidelines for permanence and
durability of the Committee on Production Guidelines
for Book Longevity of the Council on
Library Resources

Printed in the United States of America

4 6 8 10 9 7 5

Contents

Preface

Suam habet fortuna rationem

''Chance has its reason,'' says Petronius, but we may ask: what reason? and what is chance? how does chance arise? how unpredictable is the future? Physics and mathematics give some answers to these questions. The answers are modest and sometimes tentative, but worth knowing, and they are the subject of this book.

The laws of physics are deterministic. How can chance then enter the description of the universe? In several ways, as will turn out. And we shall also see that there are severe limitations on the predictability of the future. My presentation of the various aspects of chance and unpredictability will mostly follow accepted (or acceptable) scientific ideas, old and new. In particular, I shall discuss in some detail the modern ideas of chaos. The style adopted is definitely nontechnical, and the few equations that will be found in this book can be ignored without much disadvantage. High-school physics and mathematics are, in principle, all that is required to understand the main text that follows. I have, however, been less restrained in the endnotes: they range from nontechnical remarks to very technical references aimed at my professional colleagues.

Speaking of scientific colleagues, some of them will be upset by my unglorious descriptions of scientists and the world of research. For this I offer no apology: if science is the research of truth, should one not also be truthful about how science is made?

Bures-sur-Yvette
Summer 1990

Acknowledgments

While writing this book, I benefited from discussions with many colleagues. Among them I am particularly indebted to Shelly Goldstein, even though he will probably be very distressed by the text I finally wrote. Nicolas Ruelle made useful suggestions for stylistic improvement. Arthur Wightman and Laura Kang Ward fought nobly in defense of the English language. Yoshisuke Ueda and Oscar Lanford allowed reproduction of nice computer pictures. Finally, Helga Dernois showed fortitude and equanimity in typing a rather messy manuscript. Let all of them be thanked.

CHANCE AND CHAOS

CHAPTER 1

Chance

Supercomputers will some day soon start competing with mathematicians and may well put them forever out of work. At least, this is what I asserted to my very eminent colleague, the Belgian mathematician Pierre Deligne. Determined to vex him, I remarked that some machines already play chess very well, and I mentioned the proof of the four-color theorem,[1] which could be achieved only with the help of a computer. Of course, present-day machines are useful mostly for rather repetitious and somewhat stupid tasks. But there is no reason why they could not become more flexible and versatile, mimicking the intellectual processes of man, with tremendously greater speed and accuracy. In this way, within fifty or a hundred years (or maybe two hundred), not only will computers help mathematicians with their work, but we shall see them take the initiative, introduce new and fruitful definitions, make conjectures, and then obtain proofs of theorems far beyond human intellectual capabilities. Our brain, after all, has not been shaped by natural evolution with a view to perform mathematics, but rather to help us in hunting and gathering food, in making war, in maintaining social relations.

Pierre Deligne, of course, did not show great enthusiasm for my vision of the future of mathematics. After some hesitation, he told me that what interests him personally are results that he can, by himself and alone, understand in their entirety. This excludes, he said, on one hand the theorems obtained with the help of a computer, and on the other hand some extremely long mathematical proofs, resulting from the work of multiple authors, which cannot possibly be verified by a single mathematician. He was alluding to the proof of a famous theorem concerning the classification of simple finite groups.[2] This proof consists of many pieces and occupies more than five thousand pages.

Based on what I just said, one could easily paint a sinister picture of the present state of science and of its future. Indeed, if it becomes difficult for a mathematician to master a question all by himself—the proof of a single theorem—this is even more the case for his colleagues in other sciences. Whether they are physicists or physicians, in order to work efficiently, scientists use tools that they do not understand. Science is universal, but its servants are quite specialized, and their views are often limited. Undoubtedly, the intellectual and social background of scientific research has changed a lot since the origins of science. Those we now call scientists were then called philosophers, and they tried to obtain a global understanding of our world, a synthetic view of the nature of things. The great Isaac Newton characteristically shared his efforts among mathematics, physics, alchemy, theology, and the study of history in relation to the prophecies.[3] Have we given up the philosophical quest that gave birth to science?

Not at all. This philosophical quest uses new techniques but remains at the center of things. This is what I shall try to show in the present little book. There will be nothing therefore about the technical prowess of science, nothing about rockets and atom smashers. Nothing about the triumphs of medicine and nuclear dangers. No metaphysics either. I would like to don the philosophical spectacles of an honest man of the seventeenth or eighteenth century and take a walk among the scientific results of the twentieth century. A walk guided by *chance*—literally—since the study of chance will be the thread that I shall follow.

CHANCE

Chance, uncertainty, blind Fortune, are these not rather negative notions? The domain of soothsayers rather than scientists? Actually, the scientific investigation of chance is possible, and it began with the analysis of games of chance by Blaise Pascal, Pierre Fermat, Christiaan Huygens, and Jacques Bernoulli. This analysis has given birth to the *calculus of probabilities*, long considered to be a minor branch of mathematics. A central fact of the calculus of probabilities is that if a coin is tossed a large number of times, the proportion of heads (or the proportion of tails) becomes close to 50 percent. In this manner, while the result of tossing a coin once is completely uncertain, a long series of tosses produces a nearly certain result. This transition from uncertainty to near certainty when we observe *long series* of events, or *large systems*, is an essential theme in the study of chance.

Around 1900, a number of physicists and chemists still denied that matter consists of atoms and molecules. Others had long accepted the fact that in a liter of air there is an incredibly large number of molecules flying at great speed in all directions, and hitting each other in a most frightening disorder. This disorder, which has been called molecular chaos, is in effect a lot of randomness—or chance—in a little volume. How much randomness? How much chance? The question makes sense, and it is given an answer by *statistical mechanics*, a branch of physics created around 1900 by the Austrian Ludwig Boltzmann and the American J. Willard Gibbs. The amount of chance present in a liter of air or a kilogram of lead at a certain temperature is measured by the *entropy* of this liter of air or kilogram of lead, and we now have methods for determining these entropies with precision. We see that chance can thus be tamed and becomes essential in understanding matter.

You might think that what happens *at random* or *by chance* has therefore no meaning. A little bit of thinking shows that such is not the case: blood types are distributed at random in a given population, but it is not without meaning to be A+ or O− in the event of a transfusion. *Information theory*, created by the American mathematician Claude Shannon in the late 1940s, allows us to measure the information of messages that have, in principle, a meaning. As we shall see, the mean information of a message is defined as the amount of chance (or randomness) present in a set of possible messages. To see that this is a natural definition, note that by choosing a message, one destroys the randomness present in the variety of possible messages. Information theory is thus concerned, as is statistical mechanics, with measuring amounts of randomness. The two theories are therefore closely related.

Speaking of meaningful messages, I want to mention some messages that carry particularly vital information: the genetic messages. It is now a well-established fact that the hereditary characteristics of animals and plants are transmitted by the DNA in chromosomes. This DNA (deoxyribonucleic acid) is also present in bacteria and viruses (it is replaced in some viruses by ribonucleic acid). It has been shown that DNA consists of a long chain of elements belonging to four types, which may be represented by the letters A,T,G,C. Heredity therefore consists of long messages written with a four-letter alphabet. When cells divide, these messages are copied, with a few errors made at random; these errors are called *mutations*. The new cells, or the new individuals, are thus a little different from their ancestors, and more or less able to survive and reproduce. Natural selection then preserves some individuals and discards those who are less fit, or less lucky. The fundamental questions concerning life may thus be described in terms of creation and transmission of genetic messages in the presence of chance. The great problems of the origin of life and of the evolution of species are not thereby solved, but by expressing these problems in terms of creation and transmission of information we shall arrive at suggestive viewpoints and reach some quite definite conclusions.

But before investigating the creative role of chance in life processes, I should like to take you, reader, for a fairly long walk through other problems. We shall discuss statistical mechanics and information theory; we shall talk about turbulence, chaos, and the role of chance in quantum mechanics and the theory of games. We shall digress on historical determinism, black holes, algorithmic complexity, and so on.

This long promenade of ours will follow the borderline between two large intellectual territories: the austere mathematics on one side, and on the other side physics in the widest sense, including all of the natural sciences. And it will be interesting also to keep an eye on the functioning of the human mind, or brain, in its valiant and often pathetic efforts to comprehend the nature of things. Beyond the problem of chance, then, we shall try to understand something of the triangular relation between the strangeness of mathematics, the strangeness of the physical world, and the strangeness of our own human mind. For a start, I should like to discuss some rules of the games of mathematics and physics.

CHAPTER 2

Mathematics and Physics

Mathematical talent often develops at an early age. This is a common observation, to which the great Russian mathematician Andrei N. Kolmogorov added a curious suggestion. He claimed that the normal psychological development of a person is halted at precisely the time when mathematical talent sets in. In this manner, Kolmogorov attributed to himself a mental age of twelve. He gave only an age of eight to his compatriot Ivan M. Vinogradov, who was for a long time a powerful and very much feared member of the Soviet Academy of Sciences. The eight years of Academician Vinogradov corresponded, according to Kolmogorov, to the age when little boys tear off the wings of butterflies and attach old cans to the tails of cats.

Probably it would not be too hard to find counterexamples to Kolmogorov's theory,[1] but it does seem to be right remarkably often. The extreme case of a colleague comes to mind: his mental age must be something like six, and this creates some practical problems, for instance when he has to travel alone. This colleague functions relatively well as a mathematician, but I don't think that he could survive in the rather more aggressive community of physicists.

What is it that makes mathematics so special, so different from other fields of science? The starting point of a mathematical theory consists of a few *basic assertions* on a certain number of *mathematical objects* (instead of mathematical objects, we might speak of words or phrases, because in a sense that is what they are). Starting from the basic assumptions one tries, by pure logic, to deduce new assertions, called *theorems*. The words used in mathematics may be familiar, such as "point" and "space," but it is important not to trust ordinary intuition too much when doing mathematics, and to use in fact only the basic assertions given at the start. It would be entirely acceptable if you decided to say "chair" and "table" instead of point and space, and it might even be a good idea in some cases; mathematicians are not averse to making this kind of translation. If you like, then, mathematical work is something like a grammatical exercise with extremely strict rules. Starting from the chosen basic assertions the mathematician constructs a chain of further assertions, until one is generated that looks particularly nice. The mathematician's colleagues will then be invited to see the newly generated assertion, and they will admire it and say, "This is a beautiful theorem." The chain of intermediate assertions constitutes the proof of the theorem, and a theorem that can be stated simply and concisely often requires an extraordinarily long proof. The *length of proofs* is what makes mathematics interesting, and has in fact fundamental philosophical importance. To this question of length of proofs are related the problem of algorithmic complexity, and Gödel's theorem, both of which will be discussed in later chapters.[2]

Because mathematical proofs are long, they are also difficult to invent. One has to construct, without making any mistakes, long chains of assertions, and see what one is doing, see where one is going. To *see* means to be able to guess what is true and what is false, what is useful and what is not. To see means to have a feeling for which definitions one should introduce, and what the key assertions are that will allow one to develop a theory in a natural manner.

And you should not think that the mathematical game is

arbitrary and gratuitous. The diverse mathematical theories have many relations with each other: the objects of one theory may find an interpretation in another theory, and this will lead to new and fruitful viewpoints. Mathematics has deep unity. More than a collection of separate theories such as set theory, topology, and algebra, each with its own basic assumptions, mathematics is a unified whole. Mathematics is a great kingdom, and that kingdom belongs to those who see. The *seers*, those who possess mathematical intuition, derive from their power a sense of immense superiority over their blind contemporaries. They feel about nonmathematicians as jet pilots feel about ground personnel, or as the British felt in the old days about people on the Continent.

Mathematics is a kind of intellectual yoga, demanding, rigorous, ascetic. And a mathematician, a real one, invests a lot in his art. Alien concepts and strange relations occupy his thoughts, verbal or nonverbal, conscious or not. (The unconscious is often seen to play a role in mathematical discovery; a beautiful example of this was described by Henri Poincaré.)[3] The invasion of the mind by the blooming of mathematical thought, and the strangeness of this thought, put the mathematician a little apart from the rest of mankind, and one can understand that (as Kolmogorov suggested) his psychological development seems at times to have been arrested.

And what about physicists? Mathematicians and physicists often behave like enemy brothers, and tend to exaggerate their differences. But mathematics is the language of physics, as already noted by Galileo,[4] and a theoretical physicist is always to some extent a mathematician. Indeed, Archimedes, Newton, and many others have brilliantly contributed both to physics and to mathematics. The truth is that physics is intimately related to mathematics, but also profoundly different. Let me try to explain this now.

The purpose of physics is to make sense out of the world

around us. Typically, if you are a physicist, you will not try to understand everything at the same time. Rather, you will look at different *pieces of reality* one by one. You will *idealize* a given piece of reality, and try to describe it by a mathematical theory. Thus you begin by choosing a certain class of phenomena, and you define physical concepts *operationally* for that class. The physical framework being specified in this way, you still have to choose a mathematical theory and establish a correspondence between the objects of this mathematical theory and the physical concepts.[5] It is this correspondence that constitutes a *physical theory*. In principle, of course, a physical theory is better when the correspondence that it yields between physical and mathematical quantities is more precise, and when the set of phenomena that it describes is wider. But in practice the manageability of the mathematics is also important, and physicists will normally use a theory that is simple and convenient for a given application when the alternatives would be more messy and not really more accurate.

It is good to realize that the operational definition of a physical concept is not a formal definition. As our understanding progresses we can further analyze the operational definitions, but they remain less precise than the mathematical theory to which they are related. For instance, when describing chemistry experiments, you will want to specify reagents that are *reasonably pure*, and in some cases you may refine this requirement and place severe limits on the amounts of contaminants that have disastrous catalytic effects. But if you insisted on knowing beforehand the precise amount of every conceivable contaminant, you would never do any experiments. If you study physics you soon must face that apparent paradox: your control over a physical object that you can hold in your hand is less than your control over a mathematical object without material existence. This is very irritating to some people, and is in fact an essential reason why they choose to become mathematicians rather than physicists.

One modest example of a physical theory is what I would call the *theory of the game of dice*. The piece of reality that one tries to understand is what one observes when playing dice. An operationally defined concept in the theory of the game of dice is that of *independence*: one says that successive throws are independent if the dice have been well shaken between the throws. And here is an example of a prediction of the theory: for a large number of independent throws of two dice, the result will be 3 (i.e., 1 for one die, and 2 for the other) in about one case out of eighteen.

Let us summarize. By gluing a mathematical theory on a piece of physical reality we obtain a physical theory. There exist many such theories, covering a great diversity of phenomena. And for a given phenomenon there are usually several different theories. In the better cases one passes from one theory to another one by an *approximation* (usually an uncontrolled approximation). In other cases the correspondence between different physical theories leads to serious conceptual headaches because they are based on discordant and apparently irreconcilable physical concepts. In any case, jumping from one theory to another is an important part of the art of doing physics. The professionals will say that they look at *quantum corrections* or at the *nonrelativistic limit*, or they will say nothing at all because the point of view adopted is "clear from context." Under such conditions, the physical discourse may often sound a bit incoherent, if not totally confused. How do the physicists find their way in such a mess?

To answer this question you have to remember that physics has fundamental unity because it describes the unique physical universe in which we live. The unity of mathematics is due to the logical relation between different mathematical theories. The physical theories, by contrast, need not be logically coherent; they have unity because they describe the same physical reality. Physicists normally do not have existential doubts about the reality that they try to describe. Often they will need several logically incompatible theories to cover the description of a certain class of phenomena. They will of course lament this incoherence, but will not go so far as to throw away one or the other of the incompatible theories. They will keep them at least until they have found a better theory that accounts for all observed facts in a unified manner.

A last word of caution. Don't embark in general abstract discussions as to whether physics is deterministic, or probabilistic, local or not, and so on. The answer depends on the physical theory considered, and how determinism, or chance, or locality, is introduced in this theory. A meaningful physical discussion always requires an operational background. Either this is provided by an existing theory, or you have to give it yourself by the sufficiently explicit description of an experiment that can, at least in principle, be performed.

CHAPTER 3

Probabilities

The scientific interpretation of chance begins when we introduce *probabilities*. The physical concept of probability seems to be a clear and basic intuitive notion, but this does not mean that it is easily codified and formalized. As always in going from intuition to science we have to proceed with great care and caution. Let us look more closely into the problem.

"There are nine chances out of ten that it will rain this afternoon, and therefore I will take my umbrella." This kind of argument involving a probability is of constant use when we make decisions. The probability that it will rain is estimated to be 9/10, or 90 percent, or .9. Generally speaking, probabilities are counted from zero to one hundred percent or, in more mathematical terms, from 0 to 1. The probability 0 (zero percent) corresponds to impossibility and the probability 1 (one hundred percent) to certainty. If the probability of an event is neither 0 nor 1, then this event is uncertain, but our uncertainty about it is not complete. For instance, an event with probability 0.000001 (one chance in a million) is a rather unlikely event.

The success of what we undertake depends on circumstances, some of which are certain, some of which are not. It is therefore important that we assess correctly the probability of uncertain circumstances, and to do this we need a *physical theory of probabilities*. I insist that it be a *physical* theory, because it is not enough just to be able to compute probabilities, we also have to be able to compare our results *operationally* with physical reality. If we do not pay sufficient attention to the problem of relation to physical reality, we can easily get trapped in paradoxes. We should thus be a bit careful with statements like "the probability that it will rain this afternoon is .9." The operational meaning of this assertion is not clear, to say the least, and its status at this point is therefore a bit doubtful.

Consider the assertion: "When I throw a coin up in the air, the probability that it will land with heads up is .5." This sounds eminently reasonable, at least before I throw the coin, but it is obviously false when the coin has fallen, because the uncertainty is then lifted. At what time does the coin decide to show heads or tails? Suppose that you accept the principle of classical determinism, and therefore that the state of the universe at one time determines its state at any later time. Then the side on which the coin will fall is determined at the moment of the creation of the universe! Does this mean that we have to abandon probabilities, or that we may mention them only if we replace classical theory by quantum theory? *No!* That is not the way to do physics. The reasonable attitude is to introduce probabilities in a very unrestrictive framework, without speaking of classical or quantum mechanics. After specifying our concepts mathematically and operationally, we shall be in a better position to discuss the relation of probabilities to determinism, quantum mechanics, and so on.

The philosophical position that I want to defend regarding the introduction of probabilities is thus the following. For various classes of phenomena (what I called "pieces of reality" earlier) there are idealizations involving probabilities. These idealizations are of interest because they are useful: it may help to know that when you toss a coin it will show heads or tails with equal probability. It may help to know that if you toss a coin 20 times the probability of having heads every time is less than one part in a million. Estimating a probability replaces unspecified "chance" with something a bit more substantial. Our next task is to give this *something* a logically and operationally coherent structure.

If you are not familiar with probability theory (or with hard science in general), you may find the rest of this chapter a bit forbidding. Don't skip it, though! What I want to do is sketch an example of a physical theory: operationally defined physical concepts, mathematical theory, and relation between the physical and mathematical concepts. The physical theory of probabilities is what I want to describe. It is, by any standards, a very simple physical theory.

Probability theory is the art of playing with assertions like

$$\text{proba}(``A") = .9,$$

which means that the probability of the event "*A*" is 90 percent. From the mathematical viewpoint, the event "*A*" is just a symbol that should be manipulated according to certain rules. Within the framework of a physical idealization, event "*A*" is really an event, such as "it will rain this afternoon," and this has to be specified operationally. (For instance, I may decide that I shall go for a walk this afternoon, and if it rains I'll notice it. As is usually the case in physics, this operational definition is somewhat imprecise: it could happen that I am run over by a truck before the end of the afternoon, and this would put an end to my meteorological observations.)

The event "non *A*" is, from a mathematical viewpoint, simply a new assembly of symbols. In all physical idealizations that we shall want to consider, the event "non *A*" corresponds to the fact that event "*A*" does not occur. In the above example "non *A*" means "it will not rain this afternoon."

Let us now introduce, besides "*A*," a new event "*B*." From a mathematical viewpoint, this allows us to introduce new assemblies of symbols, viz. "*A* or *B*" and "*A* and *B*." These new assemblies of symbols are again events. In a physical idealization, "B" could for instance mean "there will be no rain, but there will be snow this afternoon" or "the slice of bread that I am dropping will land with the buttered side down." The event "A or B" corresponds physically to "A" happening, or "B" happening, or both "A" and "B" happening. The event "A and B" corresponds to both "A" and "B" happening.

We may now complete our mathematical presentation of probabilities by listing three basic assertions, or rules:

(1) proba("non A") = 1 − proba("A")
(2) if "A" and "B" are *incompatible*,
then proba("A or B") = proba("A") + proba("B")
(3) if "A" and "B" are *independent*,
then proba("A and B") = proba("A") × proba ("B") .

We shall come back in a minute to the discussion of these three rules, but let us remark that they involve the new and undefined concepts of *incompatible* events and *independent* events. In a treatise on probabilities, some rules would be introduced at this point on how to manipulate *non*, *and*, and *or* and the mathematical concepts of incompatible and independent events. One would also add a couple of basic assertions concerning infinite sets of events. These are important points all right, but not essential for what we plan to do, and we shall skip them.

We have just dispatched—summarily but not incorrectly—the mathematical foundations of the calculus of probabilities.[1] There remains now the equally important task of specifying the physical framework of probabilities. Or rather, the various physical frameworks, because probabilities occur in fairly diverse situations, and operational definitions have to be made case by case. Here we shall satisfy ourselves with general indications.

In physical idealizations, two events are said to be *incompatible* if they cannot occur together. Suppose that the events "*A*" and "*B*" are, respectively, "it will rain this afternoon" and "there will be no rain, but there will be snow this afternoon." Then "*A*" and "*B*" are incompatible, and rule (2) says that their probabilities add up: a 90 percent chance of rain plus a 5 percent chance of snow without rain gives a 95 percent chance of rain or snow. This is intuitively satisfactory.

Two events are said to be *independent* if they have "nothing to do" with each other, i.e., if the fact that one is realized has in the average no influence on the realization of the other. Suppose that the events "*A*" and "*B*" are, respectively, "it will rain this afternoon" and "the slice of bread that I am dropping will land with the buttered side down." I figure that these two events have nothing to do with each other, that they are unrelated, independent. By application of rule (3), their probabilities have to be multiplied: probability .9 of having rain times probability .5 of smearing butter on the floor gives a probability .45 of both events occuring. This is intuitively satisfactory: there is a 90 percent chance of rain, and half of the time the slice of bread will fall with the buttered side down, which makes a probability of 45 percent of having both rain outside and butter on the floor.[2]

We have thus verified that rules (2) and (3) are intuitively reasonable. As for rule (1), it simply says that if the probability of rain is 90 percent, the probability of no rain is 10 percent, which is hardly objectionable.

Among the notions that we have just been discussing, the concept of independence is clearly the most delicate. Experience and common sense suggest that some events are independent, but there are occasional surprises. One should therefore verify that the probabilities of supposedly independent events behave as asserted in rule (3). And the operational definitions should also be very carefully respected. Thus when playing dice, one should shake the dice well between successive throws. Only then may these throws be considered independent.

Very well, we now know how to play with probabilities, but we do not yet know to what they correspond operationally! Here then is the way to determine the probability of "*A*": perform a large number of independent experiments under conditions such that "*A*" may take place, and then observe in what proportion of the cases "*A*" actually occurs; this proportion is the probability of "*A*." (For a mathematician, a "large number" of experiments means a number that one lets tend to infinity.) For instance, if you toss a coin a large number of times, it will show heads in about half of the cases, which corresponds to a probability of .5.

Now that we have our beautiful operational definition, we may ask what is meant by the probability of the event "it will rain this afternoon." Indeed, it appears difficult to repeat "this afternoon" independently a large number of times. Some purists will therefore say that the probability in question is meaningless. One might, however, be able to give it a meaning, for instance by making a large number of numerical simulations on a computer (compatible with our present knowledge of the meteorological situation) and finding the proportion of cases in which the simulation gives rain. If one finds a probability of 90 percent for rain, even the purists will take their umbrellas.

CHAPTER 4

Lotteries and Horoscopes

I introduced probabilities in the last chapter, with mathematical basic rules, operational definitions, and so on . . . and you may wonder if all those precautions were really necessary. After all, what I said can be summarized in a few words: probabilities of incompatible events add up (to give the probability of the *or* event), probabilities of independent events multiply (to give the probability of the *and* event), and the proportion of cases in which an event occurs (in a large number of independent trials) is the probability of that event. A bit of thinking makes this all rather clear, and the subject should not give rise to controversy. When one sees, however, the success of lotteries and horoscopes (among other things), one measures how much the behavior of many people differs, with regard to probabilities, from what sound scientific thinking would dictate.

Lotteries are a freely accepted form of taxation of the less privileged layers of society. The lottery ticket that you buy, rather cheaply perhaps, is a little bit of hope of becoming rich. But the probability of hitting the jackpot is very low: it is the kind of low probability (like that of being hit by a falling object while walking in the street) that you would normally disregard. In fact, the gains, small or large, do not on average compensate for the price of the tickets, and the calculus of probabilities shows that you are practically certain to lose money if you play regularly. Let us look at the example of a somewhat simplified lottery, in which the probability of winning is 10 percent and you win 5 times the price of a ticket. Over a large number of drawings, the proportion of winnings is close to 1/10, and since you win 5 times the price of a ticket, it follows that your total gain is about half the total expenditure. The net gain is thus negative: you lose about half the money you spend. In conclusion, the more tickets you buy, the more money you lose, and this remains true for more complicated lotteries, since all of them are designed to suck money out of the player, for the benefit of the organizer.[1]

I would now like to talk about horoscopes, and I shall need for that purpose an assertion of the calculus of probabilities, which is in fact just a reformulation of rule (3) of the last chapter. Here is the assertion:

(4) if "*A*" and "*B*" are independent,
then proba("*B*," knowing that "*A*" is realized) = proba("*B*").

In other words, knowing that event "*A*" is realized tells us nothing at all about "*B*," and the probability of the latter event remains equal to proba("*B*"). This is true under the assumption that "*A*" and "*B*" are independent. (If the events "*A*" and "*B*" are not independent, one says that there are *correlations* between them, or that they are correlated.) Rule (4) is justified, for the benefit of the interested reader, in a note.[2]

We may now discuss the problem of horoscopes, which are more subtle and more interesting than lotteries because we do not see here immediately what role the probabilities play. Typically the horoscope informs you that if you are a Leo, the configuration of planets is favorable to you this week, and you will be lucky in love and at games, but if you are a Pisces you must at all costs avoid plane trips, stay home, and take good care of your health. Astronomers and physicists would object that "*X* is a Leo'' and ''*X* will win at games this week'' are independent events. Similarly for ''*X* is a Pisces'' and ''*X* will have an accident if he or she travels by plane this week.'' In fact, it is hard to imagine more beautiful examples of events that have nothing to do with each other, and are thus independent from the point of view of probability theory. We may therefore apply rule (4) above, and conclude that the probability for *X* of winning at games is the same whether *X* is a Leo or not. Similarly, the dangers of a plane trip are the same for a Pisces as for a person of any other sign of the zodiac. In conclusion, horoscopes are perfectly useless.

Is the matter judged then without appeal? Not yet, because the supporters of astrology will deny, precisely, that ''*X* is a Leo'' and ''*X* will win at games this week'' are independent events. And they will be able to present a list of illustrious astronomers who were also astrologers: Hipparchus, Ptolemy, and Kepler, for example. The best way to close the debate is thus experimental: does one find significant statistical correlations between horoscopes and reality? The answer is negative and totally discredits astrology. It must be said, however, that the discredit of astrology among scientists has a different reason: science has changed our understanding of the universe in such a way that correlations that were conceivable in antiquity have become incompatible with our present knowledge of the structure of the universe and of the nature of physical laws. Astrology and horoscopes could fit in the science of antiquity, but they do not fit in the science of today.

The situation, however, is not very simple, and it deserves a serious discussion. Because of the forces that exist between all physical bodies (universal gravitation), we know that Venus, Mars, Jupiter, and Saturn exert some effects on our old planet Earth. It is rather clear that these effects are small, and one might suppose that their influence on the course of human affairs is nil. *This is false!* In fact, certain physical phenomena, those of meteorology for instance, exhibit a great sensitivity to perturbations, so that a tiny cause can have important effects after a while. It is thus conceivable that the presence of Venus, or any other planet, modifies the evolution of the weather, with consequences that we cannot disregard. Indeed, as we shall see later, the evidence is that whether we have rain or not this afternoon depends upon, among many other things, the gravitational influence of Venus a few weeks ago! And if we look at things carefully, we find that the same arguments that tell us that Venus has an effect on the weather prevent us from knowing precisely what this effect is. In other words, rain this afternoon and the fact that Venus is here or there remain independent events as far as our use of the theory of probabilities is concerned. All this agrees with common sense, of course, but is rather more subtle than one might naively imagine (see the discussion in a note).[3]

Let us pursue our discussion. Are there situations in which the stars and planets have a very specific influence on our affairs, leading to meaningful correlations from the point of view of the theory of probabilities? Let us imagine a somewhat crazy astronomer who would, on the basis of his observations of Venus, commit sadistic crimes; wouldn't this give interesting correlations with certain horoscopes? The suggestion is not totally absurd: the ancient Mayas, who made careful observations of the cycle of Venus, were also enthusiastic human sacrificers (they cut the chest of the victim open with a flintstone knife and tore out the heart, which they then burnt down to ashes). What this means is that the intervention of human intelligence provides a mechanism by which correlations may be introduced between "events" that have a priori nothing to do with each other. How then do we know when events are really independent?

The fact is that present-day scientists have the benefit of a rather detailed knowledge of how the universe is set up, and a good understanding of how it functions. We therefore have rather precise ideas about what correlations may or may not exist. We know, for instance, that the speed of a chemical reaction may be influenced considerably by trace impurities, but not by the phase of the moon. In case of doubt, one checks. Some otherwise unexpected correlations can be brought about by intelligent agents, but there are also limitations to what those can do.

"If you are a Leo, you will be lucky in love and at games this week." What can we say about correlations between the position of planets and the private life of *X*, reader of horoscopes? As we have seen, such correlations are not impossible to the extent that they involve an intelligent agent (Maya priest, or crazy astronomer). For the rest we may exclude them. Our forefathers populated the universe with a large number of "intelligent agents"—gods, devils, elves—of which science has made a holocaust. The gods are dead . . . and human intervention cannot improve the "luck at games" of *X*. (We set up the rules so that downright cheating is not allowed.) It appears, then, that to be a Leo and to be lucky at games this week are independent events, and a statistical study would indeed confirm this. But what about the luck of *X* in love? Here human interference is not just possible, it is practically certain, owing to the intervention of our very friend *X*, reader of horoscopes, if he or she is a bit credulous. Such indeed is human nature that to believe that we have "luck in love" this week improves our self-confidence, and therefore our luck.

The unavoidable conclusion is that we often make irrational decisions, based on fortuitous coincidences that we set up as "signs" or oracles. This irrational behavior is far from being always harmful: to avoid passing under a ladder is irrational superstition but also reasonable caution. Furthermore, as we shall see, the theory of games tells us that it is advantageous to make certain decisions erratically. And finally, it is an illusion to think that we have the capability of deciding all our actions rationally.

Still, having correct ideas about probabilities can help us avoid rather serious mistakes. It is distressing to see those people who can least afford it lose their money at lotteries and similar games. Turning to horoscopes, I shall admit that I enjoy reading them once in a while. There is something almost poetical in the prediction of faraway trips, romantic encounters, or fabulous heritages . . . and those prophecies are rather harmless as long as you don't believe them too much. One may, however, be indignant at the fact that some business enterprises make hiring decisions on the basis of horoscopes. This sort of "astral" discrimination is worse than simply stupid, it is downright dishonest.

CHAPTER 5

Classical Determinism

The flow of *time* is an essential aspect of our perception of the world. And we have seen that *chance* is another essential aspect of our perception of the world. How do these two aspects fit together? Before tossing a coin, I estimate the probabilities of getting heads or tails to be both equal to 50 percent. Then, I toss the coin and get heads, say. At what moment does the coin decide to show heads? We have already asked ourselves this question, and the answer is not very easy: we are here confronted with one of those "pieces of reality" described by several different physical theories, and the connection between these different theories is a bit laborious. We discussed earlier the theory describing chance—the physical theory of probabilities. For the description of time, things get somewhat more complicated because we have at least two different theories at our disposal: *classical* mechanics and *quantum* mechanics.

Let us for a moment forget about tossing coins, and discuss mechanics. The ambition of mechanics—classical or quantum—is to tell us how the universe evolves over the course of time. Mechanics must therefore describe the motion of the planets around the sun, and the motion of the electrons around the nucleus of an atom. But while for large objects the classical theory gives excellent results, it becomes inadequate at the level of atoms and must be replaced by quantum theory. Quantum mechanics is thus more correct than classical mechanics, but its use is more delicate and difficult. And in fact neither classical nor quantum theory applies to objects with a velocity close to that of light; in such cases we have to use Einstein's relativity (either special relativity, or general relativity if we also want to describe gravitation).

But, you may tell me, why stop at classical or quantum mechanics? Don't we rather want to use the *true* mechanics, taking into account all quantum and relativistic effects? After all, what interests us is the universe as it really exists, rather than this or that classical or quantum idealization. Let us have a good look at this important question. First of all we have to face the fact that the *true mechanics* is not at our disposal. At the time of writing we do not have a unified theory that agrees with all that we know about the physical world (relativity, quanta, properties of elementary particles, and gravitation). Every physicist hopes to see such a unified theory in action, and this may happen some day, but now it is only a hope. Even if one of the theories already proposed turns out later to be the right one, it is not at this time *in action* in the sense of giving us computational access to the masses of the elementary particles, their interactions, and so on. The best we can do at present is to use a somewhat approximate mechanics. In the present chapter we shall use classical mechanics. Later we shall see that quantum mechanics is based on somewhat less intuitive physical concepts. The relation between quantum mechanics and chance will therefore be more difficult to analyze. Everything seems to indicate that the physical concepts of the true mechanics will be difficult to grasp intuitively. It is thus reasonable to use classical mechanics—with its well-known physical concepts—to investigate the relation between chance and time.

As I have just said, the ambition of mechanics is to tell us how the universe evolves over the course of time. Among other things, mechanics must describe the revolution of the planets around the sun, or the trajectory of a space vehicle powered by rockets, or the flow of a viscous fluid. In short, mechanics must describe the *time evolution* of physical systems. Newton is the first person who understood well how to do this. Using a more modern language than that used by Newton, let us say that the *state* of a physical system at a certain time is given by the positions and velocities of the points at which the matter of the system is concentrated. We must therefore give the positions and velocities of the planets, or of the space vehicle in which we are interested, or of all the points constituting a viscous fluid in the process of flowing. (In this last case there is an infinite number of points, and therefore an infinite number of positions and velocities to consider.)

According to Newtonian mechanics, when we know the state of a physical system (positions and velocities) at a given time—let us call this the initial time—then we know its state at any other time. How is this knowledge obtained? A new concept is needed here, that of *forces* acting on a system. For a given system, the forces are at each instant of time determined by the state of the system at this instant. For instance, the force of gravity between two celestial bodies is inversely proportional to the square of the distance between these bodies. Newton now tells us how the variation of the state of a system in the course of time is related to the forces acting on this system. (This relation is expressed by Newton's equation.)[1] Knowing the initial state of a system, we may then determine how this state varies in the course of time and therefore find out, as announced, the state of the system at any other moment.

I have just presented in a few words that great monument of universal thought which is Newton's mechanics, now also called classical mechanics. A serious study of classical mechanics would require mathematical tools that cannot be presented here. But some interesting remarks can be made on Newton's theory without entering into a detailed mathematical discussion. First, let us note that Newton's ideas shocked many of his contemporaries. René Descartes, in particular, could not accept the notion of "forces at a distance" between celestial bodies. He felt this idea to be absurd and irrational. Physics, according to Newton, consisted in gluing a mathematical theory on a piece of reality, and reproducing in this manner the observed facts. But this approach was too loose for Descartes. He would have wanted a *mechanistic* explanation, allowing contact forces, like that of a cogwheel on another cogwheel, but not forces at a distance. The evolution of physics has shown that Newton was right, rather than Descartes. And what would the latter have thought about quantum mechanics, in which the position and velocity of a particle cannot be simultaneously specified?

Coming back to Newtonian mechanics, we see that it gives a completely deterministic picture of the world: if we know the state of the universe at some (arbitrarily chosen) initial time, we should be able to determine its state at any other time. Laplace (or Pierre Simon, Marquis de Laplace, if you prefer) has given an elegant and famous formulation of determinism. Here it is.[2]

> An intelligence which, at a given instant, would know all the forces by which Nature is animated, and the respective situation of all the elements of which it is composed, if furthermore it were vast enough to submit all these data to analysis, would in the same formula encompass the motions of the largest bodies of the universe, and those of the most minute atom: nothing for it would be uncertain, and the future as well as the past would be present to its eyes. The human mind, in the perfection that it has been able to give to astronomy, provides a feeble semblance of this intelligence.

This quotation of Laplace has an almost theological flavor, and certainly suggests various questions. Is determinism compatible with man's free will? Is it compatible with chance? Let us first discuss chance, and then we shall have a brief look at the messy problem of free will.

At first sight, Laplace's determinism leaves no room for chance. If I toss a coin, sending it up in the air, the laws of classical mechanics determine with certainty how it will fall, showing heads or tails. Since chance and probabilities play in practice an important role in our understanding of nature, we may be tempted to reject determinism. Actually, however, as I would like to argue, the dilemma of chance versus determinism is largely a false problem. Let me try to indicate here briefly how to escape it, leaving a more detailed study for later chapters.

The first thing to note is that there is no logical incompatibility between chance and determinism. Indeed, the state of a system at the initial time, instead of being precisely fixed, may be random. To use more technical language, the initial state of our system may have a certain *probability distribution*. If such is the case, the system will also be random at any other time, and its randomness will be described by a new probability distribution, and the latter can be deduced deterministically by using the laws of mechanics. In practice, the state of a system at the initial time is never known with perfect precision: we have to allow a little bit of randomness in this initial state. We shall see that a little bit of initial randomness can give a lot of randomness (or a lot of indeterminacy) at a later time. So we see that in practice, determinism does not exclude chance. All we can say is that we can present classical mechanics—if we so desire—without ever mentioning chance or randomness. Later we shall see that this is not true for quantum mechanics. Two idealizations of physical reality may thus be conceptually quite different, even if their predictions are practically identical for a large class of phenomena.

The relations between chance and determinism have been

the object of many discussions and recently of a heated controversy between René Thom and Ilya Prigogine.[3] The philosophical ideas of these gentlemen are indeed in violent conflict. But it is interesting to note that when one comes to the specifics of observable phenomena, there is no disagreement between serious scientists. (The opposite would have been perhaps more interesting.) Let us note Thom's assertion that since the business of science is to formulate laws, a scientific study of the time evolution of the universe will necessarily produce a deterministic formulation. This need not be Laplace's determinism, however. We might very well obtain deterministic laws governing some probability distributions; chance and randomness are not so easily escaped! But Thom's remark is important with regard to the dilemma of chance versus determinism, and the related problem of free will. What Thom tells us in effect is that this problem cannot be solved by one or another choice of mechanics, because mechanics is by essence deterministic.

The problem of *free will* is a thorny one, but it cannot be left undiscussed. Let me present here briefly the point of view defended on the subject by Erwin Schrödinger, one of the founders of quantum mechanics.[4] The role left to chance in quantum mechanics has raised the hope, as Schrödinger notes, that this mechanics would agree with our ideas on free will better than Laplace's determinism does. But such a hope, he says, is an illusion. Schrödinger first remarks that there is no real problem arising from the free will of other people: we can accept an entirely deterministic explanation of all *their* decisions. What causes difficulties is the apparent contradiction between determinism and *our* free will, introspectively characterized by the fact that *several possibilities* are open, and we engage our *responsibility* by choosing one. Introducing chance into the laws of physics does not help us in any way to resolve this contradiction. Indeed, could we say that we engage our responsibility by making a choice at random? Our freedom of choice, actu-

see the future, and then uses free will to contradict the predictions. The paradox is especially pressing in science fiction stories in which there are predictors capable of making incredibly precise forecasts. (Think of Frank Herbert's *Dune* and Isaac Asimov's *Foundation*.) How do we handle this paradox? We could abandon either determinism or free will, but there is a third possibility: we may question the ability of any predictor to do the job so well that a paradox arises. Let us note that if a predictor wants to create a paradox by violating forecasts about a certain system, then the predictor must be part of the system in question. This implies that the system is probably rather complicated. But then the accurate prediction of the future of the system is likely to require enormous computing power, and the task may easily exceed the abilities of our predictor. This is a somewhat loose argument about a loosely stated problem, but I think that it identifies the reason (or one of the reasons) why we cannot control the future. The situation is similar to that of Gödel's incompleteness theorem. There also, the consideration of a paradox leads to a proof that the truth or falseness of some assertions cannot be decided, because the task of making a decision would be impossibly long. In brief, what allows our free will to be a meaningful notion is the complexity of the universe or, more precisely, our own complexity.

CHAPTER 6

Games

Normal dice have six equivalent faces numbered from 1 to 6. In order to generate random digits, it would be convenient to have dice with 10 equivalent faces numbered from 0 to 9. Actually, there is no regular polyhedron with 10 faces, but there is one with 20 faces (the icosahedron), and we can paint the same digit on opposite faces. A cast of this icosahedral die will produce a digit from 0 to 9, and each digit has the same probability, 1/10, of appearing. Furthermore, we can arrange successive casts to be independent, and obtain in this manner a sequence of independent digits. The probabilistic theory of this game of dice allows us to compute various probabilities as discussed earlier. For instance, the probability that three successive digits add up to 2 is 6/1000.

All this is not very exciting. You may thus be surprised to learn that there are printed lists of "random numbers," i.e., random digits as above, like 7213773850327333562180647 . . . It would seem that such a list is a remarkably useless possession. In this chapter I want to make a little excursion into *game theory* and prove precisely the opposite.

Here is a game you know. I have a marble that I put (behind my back) in my right or left hand, then I show you my fists and you have to guess where the marble is. We do this a number of times, noting the results. In the end we count how many times you won or lost, and settle the difference in money, beer, or some other way. I assume that both of us try to win, and that both of us are extremely clever. If I always put the marble in the same hand or if I simply alternate, you will soon notice it and win. In fact, you will eventually see through any such mechanical strategy that I might devise. Does it mean that you must necessarily win? No! If I put the marble at random with probability 1/2 in either hand, and if my successive choices are independent, then you will make a correct guess approximately half of the time, and on the average you will neither win nor lose.

That your guess will be correct half of the time (i.e., with probability 1/2) is fairly obvious. This can be proved convincingly, by remarking that my choice of hand and your guess are *independent events*. Note that it is not good enough for me to put the marble "rather randomly" in my left or right hand. Any preference for one hand, or any correlation between successive choices, will be used against me, and you will win in the long run.

Of course, I might be clever, and induce you to make wrong choices, so that you would lose, but you could counter that easily by making your guesses at random.

Now, how do I make independent successive choices of right or left hand with probability 1/2? Well, if I have a list of random numbers, I can decide that an even digit corresponds to right hand, an odd digit to left hand, and that will do the trick. Remember, however, an essential thing: my choice of hand and your guess should be independent events. Therefore, you should have no knowledge of my list of random numbers, and I should give you no hint of the hand in which I put the marble. In particular, I should not broadcast any telepathic message that you might use for your guesses. As to this last point, experiments have been made (of precisely the type of game that we are discussing), and they definitely suggest that telepathy does not exist.

So, a private list of random digits is a useful possession after all. Admittedly, the problem of acquiring a list of random digits requires further discussion, but we shall spend enough time on that later. Let us, for the moment, look at games a bit more.

The usefulness of random behavior in games is philosophically and practically an important observation (due basically to the Frenchman Emile Borel and the Hungarian-American John von Neumann). Of course, if you cooperate with someone, it is usually good to act predictably. But in a competitive situation, the best strategy often involves random, unpredictable behavior.

Let us think of a ''game'' in which I have a choice of various moves, and you decide on your own move without knowing what I did, the outcome of the game (i.e., how much I pay you or you pay me) being decided by the two moves. For instance, my move is the choice of a hand where I put a marble, and your move is guessing a hand. If you guess right I give you $1, and if you miss you give me something ($1, or whatever).

Another game would have me on a battlefield hiding in a shelter, and you circling in a little airplane, dropping bombs and trying to hit me. A natural idea is for me to select the best shelter around and hide there. But a natural idea for you is to find the best shelter and bomb it . . . so would it not be better for me to hide in the second-best shelter? If we are very clever, we shall both use probabilistic strategies. I shall compute, for the probabilities of hiding in the various available shelters, the values that give me the best overall chance of survival, and then I shall flip a coin (or use a table of random numbers) to decide where to hide. You will similarly appeal to chance to find out where to dump your bombs with the greatest overall chance of hitting me. This may sound crazy, but this is what we shall do if we both are very clever and act ''rationally.'' Of course, you can do better if I don't keep my moves secret, and you should by all means prevent me from learning your bombing intentions.

In everyday life, you will find that your boss, your lover, or your government often try to manipulate you. They propose to you a "game" in the form of a choice in which one of the alternatives appears definitely preferable. Having chosen this alternative, you are faced with a new game, and very soon you find that your reasonable choices have brought you to something you never wanted: you are trapped. To avoid this, remember that acting a bit erratically may be the best strategy. What you lose by making some suboptimal choices, you make up for by keeping greater freedom.

Of course, the idea is not just to act erratically, but to do so in accordance with a particular probabilistic strategy, involving precisely defined probabilities that we now want to compute. A specific game is determined by a table of *payoffs* as shown here.

		YOUR CHOICE			
		1	2	3	4
MY CHOICE	1	0	1	3	1
	2	−1	10	4	2
	3	7	−2	3	7

I have several choices (say 3), you have several choices (say 4), and we make our choices independently. (These choices are of the type of hiding in a certain shelter or playing a certain card in a card game.) When we have both chosen, there is a certain payoff, given by the table above. For instance, my choice 2 and your choice 4 give a payoff of two dollars, which you have to pay me. If I make the choice 3 and you the choice 2, the payoff is minus two dollars, i.e., I have to pay you two dollars.

Suppose that I make my three choices with certain probabilities and you make your four choices with certain probabilities. All these probabilities determine a certain average payoff (or expected payoff) that you will try to make minimum and I shall try to make maximum. In 1928, J. von Neumann proved that my maximum of your minimum is the same as your minimum of my maximum (this is the celebrated *minimax theorem*).[1] What this means is that, as both of us are very clever players, we agree precisely on how to disagree.

I shall not go into the detailed mathematical problem of computing the probabilities of your choices and of my choices, and the average payoff. This problem is of a general type called *linear programming*, and it is not too hard when the choices open to you and me are few. When the table of payoffs becomes large, the problem becomes more difficult. We shall later discuss precisely how difficult linear programming is.

The theory of games, as we have seen, is a nice mathematical theory showing that a secret source of random digits is a useful thing to have. But perhaps we live in a deterministic universe where nothing is random. Short of having God Almighty send us random digits on a private line, what can we do? We may roll a die or flip a coin and assert that under certain operationally defined conditions, this produces random choices. But, at some point, we shall have to find out how this randomness arises. This is a somewhat complicated affair, and it will take us the next few chapters to clarify it.

CHAPTER 7

Sensitive Dependence on Initial Condition

You remember the story of the wise man who invented the game of chess. As a reward, he asked the king to put one grain of rice on the first square of the chessboard, two grains on the second square, four on the third, and so on, doubling the number of grains of rice at each square. The king first thought that this was a very modest reward, until he found that the amount of rice needed was so huge that neither he nor any king in the world could provide it. This is easy to check: if you double a quantity ten times, you multiply it by 1024, if you do so twenty times you multiply it by more than a million, and so on.

A quantity that doubles after a certain time, and then doubles again after the same interval of time, and then again and again, is said to grow *exponentially*. As we have just seen, it will soon be huge. Exponential growth is also called *growth at constant rate*: if you put money in a bank at a constant rate of 5%, the amount will double in about 14 years provided you can ignore taxes and inflation. This type of growth is fairly natural, and occurs commonly in the real world, . . . but never lasts very long.

We shall use the idea of exponential growth to understand what happens when you try to stand a pencil in equilibrium on its point. Unless you cheat you won't succeed. This is because you never have the pencil exactly in equilibrium, and any deviation will cause the pencil to fall on one side or the other. If one studies the fall of the pencil by the laws of classical mechanics (which we won't do) one finds that it falls *exponentially fast* (approximately, and at least at the beginning of the fall). Thus the deviation of the pencil from equilibrium will be multiplied by 2 in some interval of time, then again by 2 in the next interval of time, and so on, and very soon the pencil will be lying flat on the table.

Our discussion of the pencil gives an example of *sensitive dependence on initial condition*. This means that a small change in the state of the system at time zero (the initial position or velocity of the pencil) produces a later change that grows exponentially with time. A small cause (pushing the pencil slightly right or left) has then a big effect. It would seem that for this situation to occur (small cause producing big effect), one needs an exceptional state at time zero, like the unstable equilibrium of a pencil on its point. The opposite is true: *many physical systems exhibit sensitive dependence on initial condition for arbitrary initial condition*. This is somewhat counterintuitive, and it has taken some time for mathematicians and physicists to understand well how things happen.

Let me present another example—that of a game of billiards with round or convex obstacles. As physicists always do, we idealize the system a bit: we ignore "spins," we neglect friction, and we assume that collisions are *elastic*. We are interested in the motion of the center of the billiard ball, which is rectilinear and uniform as long as there are no collisions. When there is a collision of the billiard ball with an obstacle, we think, instead, of the center of the ball being reflected by a larger obstacle (larger by precisely the radius of the ball—see Figure 7.1). The way the trajectory of the center of the billiard ball is reflected by an obstacle is exactly the same as the way a light ray is reflected by a mirror (this is what is meant by an elastic collision). With this mirror analogy we are now in a good position to discuss changes in initial condition for the billiards problem.

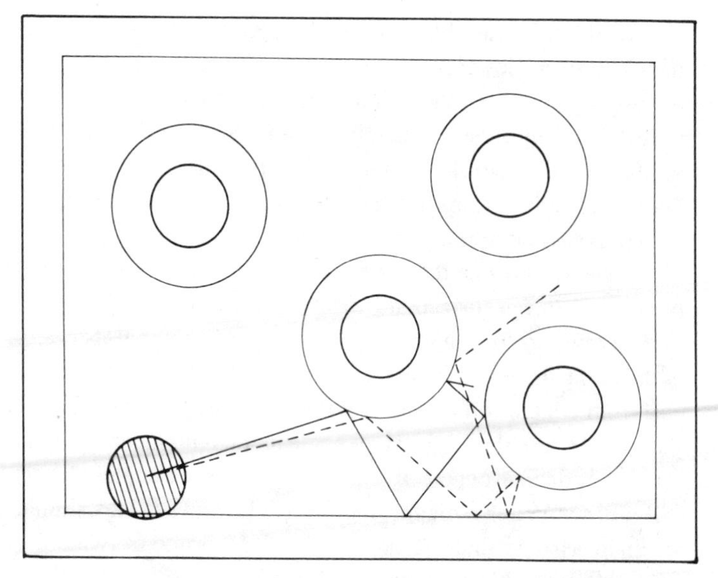

Figure 7.1. A billiard table with convex obstacles. The ball starts in the bottom left corner, and the subsequent trajectory of its center is marked (continuous line). An imaginary ball is started in a slightly different direction (dashed line). After a few collisions, the two trajectories no longer have anything to do with each other.

Suppose thus that we have on the same billiard table a *real* and an *imaginary* ball. We hit them simultaneously so that they have the same speed but slightly different directions of motion. The trajectories of the real and imaginary balls thus form a certain angle—which we shall pompously call the angle *alpha*—and the distance between the two balls increases proportionally with time. Note that this growth proportional to time is not the explosive *exponential* growth of distances that we discussed earlier. If after one second of time the centers of the real and the imaginary balls are a distance of one micron (one thousandth of a millimeter) apart, they will be only twenty microns apart (which is still very small) after twenty seconds.

A little bit of thinking shows that reflection on the straight side of the billiard table will not change the situation: the reflected trajectories form the same angle alpha as before, and the distance between the real and imaginary balls remains proportional to the time. Remember that the reflection of the ball on the side of the billiard table obeys the same laws as the reflection of light on a mirror: as long as the mirror is straight we do not expect anything very interesting to happen.

But we did say that there were round obstacles on the billiard table, and those correspond to convex mirrors. If you have looked at yourself in a convex mirror, you know that its effect is different from that of a plane mirror. What happens is discussed in courses on optics, and it is basically the following: if you send a pencil of light at an angle alpha onto a convex mirror, the reflected pencil of light has a different angle—let us call it *alpha prime*—bigger than alpha. To make things simple we shall assume that the new angle alpha prime is twice the angle alpha. (This is oversimplifying a bit, as we shall see later.)

Let us come back to our billiard table, with round obstacles, and to our two billiard balls, one real and one imaginary. Initially, the trajectories of the two balls make an angle alpha, and this is not changed by reflection on the straight sides of the billiard table. After the balls hit a round obstacle, however, their trajectories diverge, and form an angle alpha prime, which is twice the original angle alpha. Another shock on a round obstacle will give trajectories diverging at an angle equal to 4 alpha. After 10 shocks the angle is multiplied by 1024, and so on. If we have one shock per second, the angle between the trajectories of the real and the imaginary balls grows exponentially with time. In fact, it is mathematically easy to show that the distance between the two balls also grows exponentially with time as long as it remains small:[1] *we have sensitive dependence on initial condition.*

Suppose now that things are set up so that the distance between the centers of the real and the imaginary balls doubles every second. Then, after ten seconds, an original distance of one micron has increased to 1024 microns—about 1 millimeter. After twenty (or thirty) seconds the distance would have grown to more than one meter (or one kilometer)! But this is nonsense: the billiard table is not that large. The reason for this nonsense is our oversimplification in assuming that after reflection by a round obstacle the angle of the trajectories of our two billiard balls was multiplied by two but remained small. While this may be approximately correct as long as the two trajectories are close together, it fails later: the "real" trajectory will hit an obstacle and the "imaginary" trajectory will miss it entirely (or vice versa).

Let me now summarize our findings about the motion of a ball on a billiard table with round obstacles. If we observe simultaneously the motion of the "real" ball, and that of an "imaginary" ball with slightly different initial condition, we see that the two motions *usually* separate exponentially with time for a while, then one ball hits an obstacle that the other misses, and from then on the two motions no longer have anything to do with each other. To be honest, I must mention that there are exceptional initial conditions for the "imaginary" ball, such that the motions of the two balls do not separate exponentially; for instance, the imaginary ball might follow the real one on the same trajectory but one millimeter behind. This is exceptional, however, and usually the two trajectories diverge as asserted.

Before leaving the subject, let me stress that I have presented above only a *heuristic* discussion of billiards. This means that I have made things plausible, but not given a proof. It is important that one can, along the same lines, perform a completely rigorous mathematical analysis of billiards with convex obstacles. This analysis (due to the Russian mathematician

CHAPTER 8

Hadamard, Duhem, and Poincaré

I hope that I have convinced you in the preceding chapter that there is something strange going on in billiards with convex obstacles. Suppose I slightly modify the initial condition, replacing the "true" position of the ball and direction of the shot with a slightly different "imaginary" position and direction. Then the "true" and the "imaginary" trajectories, which are at first very close to each other, later diverge more and more rapidly until they no longer have anything to do with each other. This is what we have called sensitive dependence on initial condition. Conceptually, this is a very important discovery. The motion of our billiard ball is precisely determined by the initial condition, yet there is a fundamental limitation in predicting the trajectory. We have determinism, yet long-term unpredictability. This is because we know the initial condition with a certain imprecision: we cannot distinguish between the "true" initial condition and the many "imaginary" initial conditions that are very close to it. We do not know, therefore, which of the possible predictions is correct. But if the motion of a billiard ball is unpredictable, what about the motion of planets? the evolution of the weather? the fate of empires? These are interesting questions, and the answers, as we shall see, are diverse. The motion of planets is predictable over centuries, but the evolution of the weather can be forecast usefully for one or two weeks, at most. To argue about the fate of empires and the history of mankind is very ambitious indeed, yet some conclusions are possible even there and point to unpredictability. One can understand the enthusiasm of scientists when they realized that such problems were within their grasp.

Yet we have to proceed cautiously. If you have the critical mind of a scientist, you will want to clarify a few points about billiards before allowing me to embark in speculations about the predictability of the future of mankind.

For example, in studying the motion of the billiard ball, we have neglected friction. Can we do that? This kind of question arises all the time in physics: is a certain idealization allowed? Here, the presence of friction implies that the ball will eventually come to rest. But if it comes to rest long after the motion has become unpredictable, then the idealization that there is no friction is useful.

We now have to face a more serious question: how general is sensitive dependence on initial conditions? We have considered a particular system, billiards with convex obstacles, and we have argued that a little uncertainty in initial condition leads to long-term unpredictability. Are most systems like this, or is the situation exceptional? What I mean by a "system" is either a mechanical system without friction, or a system with friction but with a source of energy to replace the energy dissipated by friction, or more generally a system with electric or chemical components, etc. What is important is that we have a well-defined *deterministic* time evolution. Mathematicians then say that we have a *dynamical system*. Planets orbiting around a star form a dynamical system (an essentially frictionless mechanical system). A viscous fluid churned by a propeller is also a dynamical system ("dissipative" in this case, because there is friction). If we can find a suitable idealization of the history of mankind as a deterministic time evolution, this too is a dynamical system.

But let us come back to our question: Is sensitive dependence on initial condition the exception or the rule among dynamical systems? Do we usually have long-term predictability or not? In fact, various possibilities exist. In some cases there is no sensitive dependence on initial condition (think of a pendulum with friction, which will go to rest in a very predictable way). In other cases there is sensitive dependence on initial condition for all initial conditions (this was the case for our billiards with convex obstacles, and you will have to accept my word that this is not a completely exceptional situation). Finally, many dynamical systems are such that for some initial conditions there is long-term predictability, for others not.

On one hand it may sound a bit disappointing that all those possibilities exist. On the other hand, suppose we can tell which systems have sensitive dependence on initial condition, and for how long one can trust predictions of their future. Then we really have learned something useful about the nature of things.

It may be a good idea, at this point, to look into sensitive dependence on initial condition from a historical point of view. People of course realized thousands of years ago that small causes can have large effects, and that the future is hard to predict. What is relatively new is the demonstration that for some systems, small changes of initial condition usually lead to predictions so different, after a while, that prediction becomes in effect useless. This demonstration was made at the end of the nineteenth century by the French mathematician Jacques Hadamard[1] (who was then about thirty; he lived on to an old age, and died in 1963).

The system considered by Hadamard was a strange kind of billiards in which the flat table is replaced by a twisted *surface of negative curvature*. One studies the frictionless motion of a point on the surface. Hadamard's billiards is thus what is called in technical terms the *geodesic flow* on a surface of negative curvature. This geodesic flow is mathematically quite manageable,[2] and Hadamard could prove sensitive dependence on initial condition as a theorem. (The proof is much easier than the corresponding proof for billiards with convex obstacles, given much later by Sinai in the 1970s.)

The French physicist Pierre Duhem is one of those who understood the philosophical significance of Hadamard's result. (Duhem had ideas ahead of his time in many domains, although his political views were distinctly reactionary.) In a book for the general public published in 1906, Duhem has a section with the title "Example of a mathematical deduction forever unusable."[3] The mathematical deduction in question, as he explains, is the calculation of a trajectory on Hadamard's billiard table. It is "forever unusable" because a small uncertainty necessarily present for the initial condition will result in a large uncertainty for the predicted trajectory if we wait long enough, and this makes the prediction useless.

Another French scientist who wrote philosophical books on science at the time is the famous mathematician Henri Poincaré. In his book *Science et Méthode*, published in 1908,[4] he discusses the question of unpredictability in a very nontechnical way. He does not quote Hadamard or the relevant mathematics (but remember that Poincaré created the theory of dynamical systems, and knew more than anyone else on the subject). An essential point made by Poincaré is that *chance* and *determinism* are reconciled by long-term unpredictability. Here it is, in one crisp sentence: *A very small cause, which escapes us, determines a considerable effect which we cannot ignore, and we then say that this effect is due to chance.*

Poincaré knew well how useful probabilities are for describing the physical world. He knew that chance is part of everyday life. Since he also believed in classical determinism (there was no quantum uncertainty in his time), he wanted to understand how chance crept in. He obviously thought quite a bit about the problem, and he came up with several answers. In other words, Poincaré saw several ways in which a classical deterministic description of the world could naturally lead to a probabilistic idealization. One of these ways is through sensitive dependence on initial condition.[5]

Two examples of sensitive dependence on initial condition are discussed by Poincaré. The first is that of a gas composed of many molecules flying at great speed in all directions and undergoing many collisions. Poincaré asserts that these collisions produce sensitive dependence on initial conditions. (The situation is analogous to that of a billiard ball hitting convex obstacles.) The unpredictability of the collisions in the gas justifies a probabilistic description.

Poincaré's second example is meteorology, and he argues that the well-known unreliability of weather forecasts is due to sensitivity to initial conditions, combined with the somewhat inaccurate knowledge that we necessarily have of initial conditions. As a result, the evolution of the weather appears to be due to chance.

For a present-day specialist, the most striking thing about Poincaré's analysis is how modern it is. Both the dynamics of a gas of hard spheres and the circulation of the atmosphere have been prime objects of study in recent years using the viewpoint adopted by Poincaré.

Equally striking is the long gap in time between Poincaré and the modern study of sensitive dependence on initial condition by physicists. When related ideas surfaced again, forming what is now called the theory of *chaos*, the physical insight of Hadamard, Duhem, and Poincaré played no role in the process. Poincaré's mathematics (or what it had become) did play a role, but his ideas on weather forecasts had to be rediscovered in another way.

I think there are two reasons for this puzzling historical gap. The first is the advent of quantum mechanics. The new mechanics changed the scientific landscape of physicists and occupied all their energies for many years. Why would they bother, for instance, to explain chance by sensitive dependence on initial condition in classical mechanics when quantum mechanics introduced a new—more intrinsic—source of chance and randomness?

I see another reason why the ideas of Duhem and Poincaré fell into oblivion instead of being followed by the modern theory of chaos. These ideas came too early: the tools to exploit them did not exist. The mathematics of measure theory or the ergodic theorem, for instance, were not available to Poincaré, and therefore he could not express his brilliant intuitive ideas on chance in a precise language. When a present-day scientist reads Poincaré's philosophical writings, he has in the back of his mind a whole system of concepts by which he interprets the ideas presented, but these concepts were not available to Poincaré himself! Another fact is that when mathematics fails, we can now resort to computer simulation. This tool, which has played such an essential role in the modern theory of chaos, did not of course exist in the early twentieth century.

CHAPTER 9

Turbulence: Modes

On a rainy day in 1957, a little funeral procession carried the mortal remains of Professor Theophile De Donder to a Belgian cemetery. The hearse was accompanied by a detachment of gendarmes on horseback. The deceased had the right to this honor, and the widow had wanted it. A few unhappy scientific colleagues followed the hearse.

Theophile De Donder was the spiritual father of mathematical physics at the Free University of Brussels, and therefore one of my own spiritual grandfathers. He had, in his day, done excellent research work in thermodynamics and in general relativity (Einstein called him "le petit Docteur Gravitique").[1] But when I knew him, he was a desiccated little old man, well past doing any scientific work. Intellectual strength had left him forever, but not the desire, the fascination, which are at the root of scientific work. When he could corner some colleague at the university, he would tell the unfortunate about his research on the "ds^2 of music" or the "mathematical theory of the shape of the liver." Indeed, music, shapes, and forms are recurrent themes that fascinate scientists.[2] Here are some others: time and its irreversibility, chance and randomness, life. There is one phenomenon—the motion of fluids—that seems to reflect and combine all these themes. Think of air flowing through organ pipes, of water forming eddies and whorls perpetually changing and moving as if they had their own free will. Think of volcanic eruptions, think of wells and cascades.

CHAPTER 9

There are various ways of honoring beauty. Where an artist would paint or write a poem or compose a piece of music, a scientist creates a scientific theory. The French mathematician Jean Leray spent time, as he told me, watching the vortices and eddies in the Seine river as it flows past the piles of the Pont-Neuf in Paris. This contemplation was one of the sources of inspiration for his great 1934 paper on hydrodynamics.[3] Many great scientists have been fascinated by the motion of fluids, particularly the kind of complicated, irregular, apparently erratic motion that we call turbulence. What is turbulence? There is no obvious answer to this question, and it remains debated to this day even though people would agree, when they see a turbulent flow, that it is indeed turbulent.

Turbulence is easy to see but hard to understand. Henri Poincaré thought about hydrodynamics and taught a course on vortices,[4] but he did not risk a theory of turbulence. The German physicist Werner Heisenberg, founder of quantum mechanics, proposed a theory of turbulence that never gained much acceptance. It has been said that "turbulence is a graveyard of theories." Of course, there have been beautiful contributions to the physics and the mathematics of fluid motion by such people as Osborne Reynolds, Geoffrey I. Taylor, Theodore von Kármán, Jean Leray, Andrei N. Kolmogorov, Robert Kraichnan, and others, but the subject does not seem to have revealed its last mysteries.

In this chapter and the following ones I would like to relate one episode of the scientific struggle to understand turbulence and the later theory of chaos. This is an episode in which I myself participated, so that I can fill in more details than for events involving the half-mythical scientific giants of the beginning of the century. I shall try to give an idea of the atmosphere of research, not a balanced historical account. For the latter, it is best to refer the reader to the original papers, many of which have been conveniently collected in two reprint volumes.[5]

The discovery of new ideas cannot be programmed. This is why revolutions and other social cataclysms often have a positive influence on science. By temporarily interrupting the routine of bureaucratic chores and putting the organizers of scientific research out of commission, they give people the opportunity to think. Be that as it may; the social "events" of May 1968 in France were welcome to me because they disrupted mail and communications, and also produced some amount of intellectual excitement. I was at that time trying to teach myself hydrodynamics by reading the book *Fluid Mechanics* by Landau and Lifshitz. I worked my way slowly through the complicated calculations that these authors seem to relish, and suddenly fell on something interesting: a section on the onset of turbulence, without complicated calculations.

To understand Lev D. Landau's theory of the onset of turbulence, you must remember that a viscous fluid, such as water, will eventually go to rest unless something is done to keep it in motion. Depending on how much power is used to keep the fluid moving, you will see different things. To take a concrete example, think of water running through a tap. The power applied to the fluid (which is ultimately due to gravity) is regulated by opening the tap more or less. If you open the tap a very small amount, you can arrange a *steady* stream of water between the tap and the sink: the column of water appears motionless (although of course the tap is running). Carefully opening the tap a bit more you can (sometimes) arrange regular pulsations of the fluid column; the motion is said to be *periodic* rather than steady. If the tap is opened more, the pulsations become irregular. Finally, if the tap is wide open you see a very irregular flow; you have *turbulence*. Such a succession of events is typical for a fluid excited by a progressively greater external power source. Landau interprets this by saying that as the applied power is increased, more and more *modes* of the fluid system are excited.

We must at this point take a plunge into physics and try to understand what a mode is. Many objects around us start oscillating or vibrating when we hit them: a pendulum, a metal rod, a string of a musical instrument are readily set into periodic motion. Such a periodic motion is a *mode*. There are also modes of vibration of the column of air in an organ pipe, modes of oscillation of a suspended bridge, and so on. A given physical object often has many different modes, which we may want to determine and control. Think for instance of the design of a church bell: if the different vibration modes of the bell correspond to discordant frequencies, the sound will not be pleasant. An important example of modes is provided by the vibration of the atoms around their mean positions in a piece of solid matter; the corresponding modes are called *phonons*. But let us return to Landau. His proposal was that when a fluid is set into motion by an external power source, a certain number of modes of the fluid are excited. If no mode is excited we have a steady state of the fluid. If a single mode is excited we have periodic oscillations. If several modes are excited, the flow becomes irregular, and when many modes are excited, it is turbulent. Landau supported his proposal by mathematical arguments that I cannot reproduce here. (Independently of Landau, the German mathematician Eberhard Hopf published a similar theory, with slightly more mathematical sophistication.)[6] Turning to physical experimentation, one can make a time frequency analysis of the oscillation of a turbulent fluid, i.e., look for the frequencies that are present. One finds that many frequencies occur—in fact, a continuum of frequencies—which must therefore correspond to very many excited modes of the fluid.

As I have presented it, the Landau-Hopf theory seems to give a satisfactory description of the *onset of turbulence*: the manner in which a fluid becomes turbulent when the external power applied to it is increased. Yet, when I read Landau's explanation I was immediately dissatisfied, for mathematical reasons which will become clear in a moment.

A few more words about modes are in order at this point. In many cases you can set a physical system oscillating according to several different modes simultaneously, and the different oscillations do not influence each other. Admittedly this is not a very precise statement. To have a more definite picture, think of the modes as being oscillators somehow contained in our physical system, and oscillating independently. This mental picture has actually been very popular with physicists.

Using the terminology of Thomas Kuhn[7] we may say that the interpretation of large areas of physics in terms of modes, conceived of as independent oscillators, is a paradigm. Because of its simplicity and generality, the modes paradigm has been remarkably useful. It works whenever one can define modes that are independent or almost independent. For instance, the oscillation modes of atoms in a solid, the so-called phonons, are not quite independent: there are phonon-phonon interactions, but they are relatively small and physicists can handle them (to some extent).

The reason why I did not like Landau's description of turbulence in terms of modes is that I had heard seminars by René Thom and studied a fundamental paper by Steve Smale[8] called "Differentiable dynamical systems." The Frenchman René Thom and the American Steve Smale are both eminent mathematicians. The former is my colleague at the Institut des Hautes Etudes Scientifiques near Paris, and the latter makes frequent visits there. From them I had learned the modern developments of Poincaré's ideas on dynamical systems, and from these developments, it was clear that the applicability of the modes paradigm is far from universal. For example, a time evolution that may be described in terms of modes cannot have sensitive dependence on initial condition. I shall justify this statement in the next chapter, and show that the time evolution given by modes is rather dull as compared with the time evolutions discussed by Smale. The more I thought about the problem, the less I believed Landau's picture: if there were modes in a viscous fluid they would interact strongly rather than weakly and produce something quite different from the modes picture. Something richer and much more interesting.

Now, what does a scientist do when he thinks he has discovered something new? He produces a paper, an article, written in codified jargon, which is sent to the editor of a scientific journal, to be considered for publication. The editor uses one or more colleagues as "referees" for the paper, and if it is accepted, it is eventually printed in the scientific journal in question. Don't look for such journals at your newsstand—they are not for sale there. They are mailed to scientists, they fill the cabinets in the offices of university professors, and there are miles of them stacked in big scientific libraries.

Writing a paper entitled "On the nature of turbulence" was a joint venture with Floris Takens, a Dutch mathematician who contributed his mathematical expertise and was not afraid to stick his neck out writing a paper about physics. In the paper we explained why we thought that the Landau picture of turbulence was wrong, and we proposed something else, involving *strange attractors*. These strange attractors came from Steve Smale's paper, but the name was new, and nobody now remembers if Floris Takens invented it, or I, or someone else. We submitted our manuscript to an appropriate scientific journal, and it soon came back: rejected. The editor did not like our ideas, and referred us to his own papers so that we could learn what turbulence really was.

I shall for the time being abandon "On the nature of turbulence" to its uncertain future, and turn to something more fascinating: strange attractors.

CHAPTER 10

Turbulence: Strange Attractors

Mathematics is not just a collection of formulas and theorems; it also contains ideas. One of the most pervasive ideas in mathematics is that of *geometrization*. This means, basically, visualization of all kinds of things as points of a space.

There are many practical applications of "geometrization" in terms of charts and diagrams. Suppose that you are interested in wind-chill factors; it will be convenient for you to plot points in a temperature-airspeed diagram like Figure 10.1(a).

One advantage of this representation is that you are not committed to a single system of units. If you are a pilot, a representation like that of Fig. 10.1(b) will be useful: it gives the direction of the wind as well as its speed. It would be possible to have the direction and speed of the wind as well as the air temperature in the same three-dimensional diagram; it is easy to imagine this diagram, but only a two-dimensional projection of it can be drawn on a sheet of paper. If you also want to represent the barometric pressure and the relative humidity, you need a five-dimensional space, and you might think that a geometric picture is now impossible or useless. Has it not been said that the only people who can "see in four dimensions" are locked up in lunatic asylums? Well, the truth is that many mathematicians and other scientists routinely visualize things in 4,5, . . . or an infinite number of dimensions. One part of

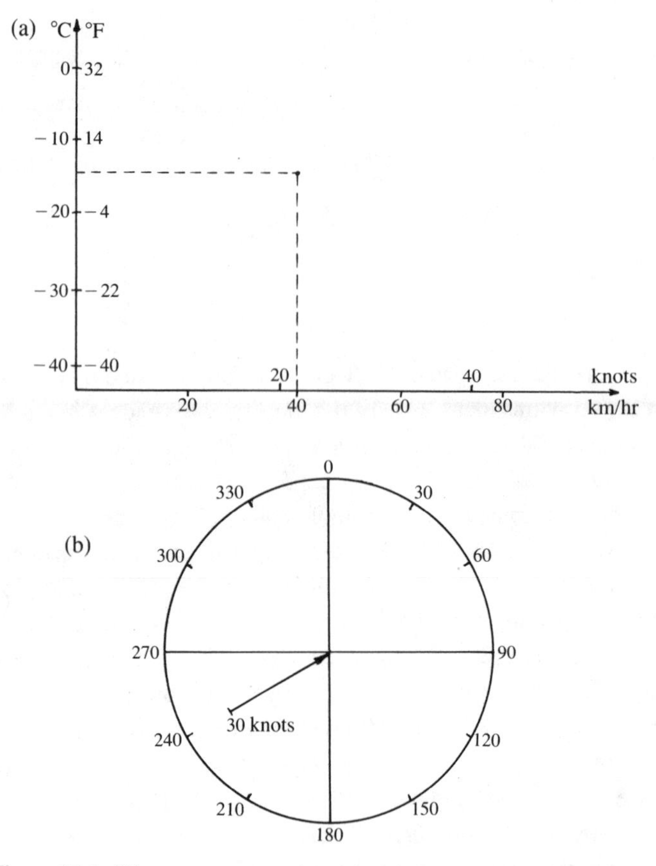

Figure 10.1. Diagrams representing (a) airspeed and temperature or (b) windspeed and direction.

the trick is to visualize various 2- or 3-dimensional projections; another part is to have a few theorems in mind that tell you how things should be. For instance, Fig. 10.2(a) is in 10 dimensions and shows a straight line intersecting in two points a 9-dimensional sphere (this 9-sphere or "hypersphere" consists of points at some fixed distance from a point O); the dotted part of the line is the part inside the sphere.

In fact, Fig. 10.2(a) represents the intersection of a straight line and a hypersphere in any number of dimensions greater than or equal to 3 (for instance, in an infinite number of dimensions). The situation in 2 dimensions is shown in Figure 10.2(b).

We shall now go back to the oscillations or "modes" of the last chapter, and try to geometrize them. The position of a pendulum, or vibrating rod, or other oscillating object, is plotted in Figure 10.3(a). This position oscillates from left (L) to right (R), then back from R to L, and so on. This picture is not very informative, but we have forgotten something: the state of our oscillating system is not quite determined by its position; we also have to know its velocity. In Figure 10.3(b) we see the orbit representing our oscillator in the position-velocity plane. This orbit is a loop (a circle if you wish), and the point representing the state of our oscillator turns around the loop with a certain periodicity.

Let us turn now to a fluid system like the running tap that we considered earlier. In our discussion we shall concentrate on the long-term behavior of the system, ignoring *transients* that occur, for instance, at the moment when we open the tap. To represent our system we need an infinite-dimensional space, because we have to specify the velocity at all points of space occupied by the fluid, and there is an infinity of such points.

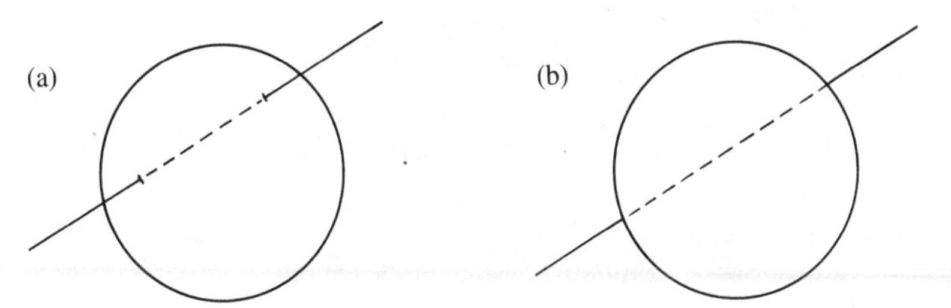

Figure 10.2. (a) A line intersecting a sphere in 10 dimensions; (b) the same thing in 2 dimensions.

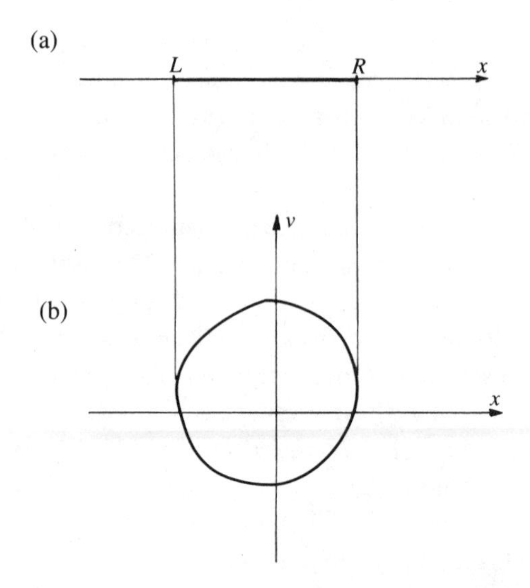

Figure 10.3. Oscillating point: (a) position x; (b) position x and velocity v.

But that need not bother us. Figure 10.4(a) shows a steady state of the fluid: the point P representing the system does not move. Figure 10.4(b) shows periodic fluid oscillations: the orbit of the representative point P is now a loop around which P circles periodically.

It is convenient to "straighten" the picture 10.4(b) so that the loop becomes a circle, and the motion on it is at constant speed. (This is done by what mathematicians call a nonlinear change of coordinates; it is like looking at the same picture through a distorting glass.) Our periodic oscillation or "mode" is now described by Fig. 10.5(a).

We have at this point all the ideas needed to visualize a superposition of several modes: as shown in Figure 10.5(b), the point P representing the system appears in several different projections to be running around circles at different angular velocities corresponding to different periods. (The projections have to be suitably chosen, and this involves nonlinear changes

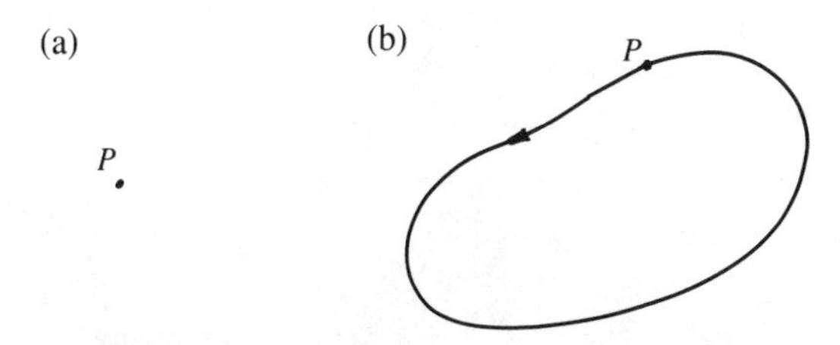

Figure 10.4. (a) Fixed point P representing a steady state; (b) periodic loop representing a periodic oscillation of a fluid. The things pictured are in an infinite number of dimensions, projected on the two-dimensional sheet of paper.

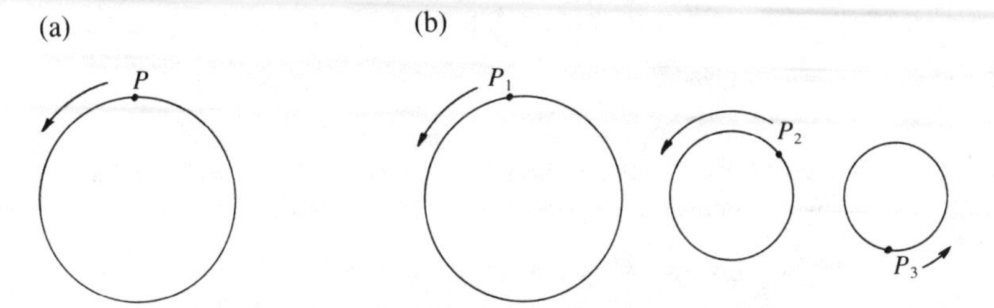

Figure 10.5. (a) Periodic oscillation (mode) described by a point P moving at constant speed around a circle; (b) a superposition of several modes described in several different projections.

of coordinates.) The interested reader may check that this time evolution does *not* have sensitive dependence on initial condition.[1]

And now look at Figure 10.6! This is a perspective view of a time evolution in three dimensions. The motion takes place on a complicated set, called a strange attractor, and more precisely the *Lorenz attractor*.[2]

Edward Lorenz is a meteorologist who worked at the Massachusetts Institute of Technology. As a meteorologist, he was interested in the phenomenon of atmospheric convection. Here is the phenomenon: the sun heats the ground, and therefore the lower layers of atmospheric air become warmer and lighter

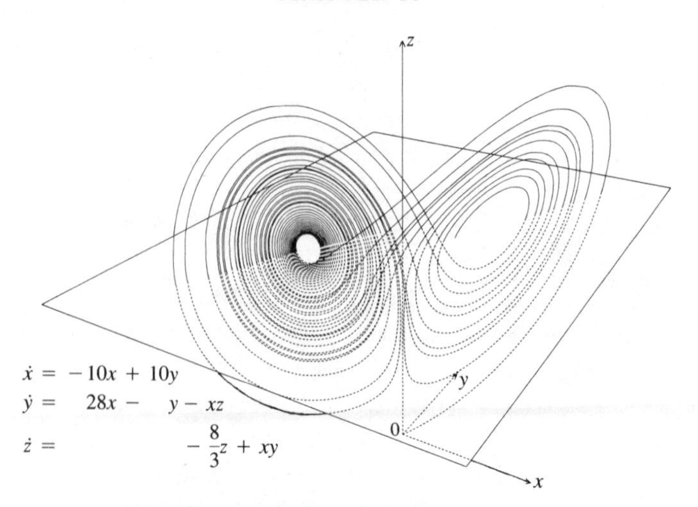

Figure 10.6. The Lorenz attractor. A computer picture programmed by Oscar Lanford (see p. 114 in *Lecture Notes in Math*. no. 615 [Berlin: Springer, 1977]; reproduced by permission).

than higher layers. This causes an upward motion of light, warm air and a downward motion of dense, cold air. These motions constitute convection. Air is a fluid like the water discussed earlier, and it should be described by a point in infinite-dimensional space. By a crude approximation, Ed Lorenz replaced the correct time evolution in infinite dimension by a time evolution in three dimensions, which he could study on a computer. What came out of the computer is the object shown in Figure 10.6, now known as the *Lorenz attractor*. We have to imagine the point P representing the state of our convecting atmosphere as moving with time along the line drawn by the computer. In the situation depicted, the point P starts near the origin O of coordinates, then turns around the right "ear" of the attractor, then a number of times around the left ear, then twice around the right ear, and so on. If the initial position of P near O were changed just a little bit (so that the difference would not be visible to the naked eye), the details of Figure 10.6 would be completely changed. The general aspect would remain the same, but the number of successive turns around the right and left ears would be quite different. This is because—as Lorenz recognized—the time evolution of Figure 10.6 has sensitive dependence on initial condition. The number of successive turns around the left and right ears is thus erratic, apparently random, and difficult to predict.

The Lorenz time evolution is not a realistic description of atmospheric convection, but its study nevertheless gave a very strong argument in favor of unpredictability of the motions of the atmosphere. As a meteorologist, Lorenz could thus present a valid excuse for the inability of his profession to produce reliable long-term weather predictions. As we have seen, Poincaré had made precisely the same remark much earlier (Lorenz was not aware of this). But the approach of Lorenz has the great virtue of being specific, and extendable to realistic studies of the motion of the atmosphere. Before leaving Lorenz, let me note that while his work was known to meteorologists, physicists became aware of it only rather late.

Let me now come back to the paper "On the nature of turbulence" which I wrote with Floris Takens and which we abandoned in the last chapter. The paper was eventually published in a scientific journal.[3] (Actually, I was an editor of the journal, and I accepted the paper myself for publication. This is not a recommended procedure in general, but I felt that it was justified in this particular case.) "On the nature of turbulence" has some of the same ideas that Poincaré and Lorenz developed earlier (we were unaware of this). But we were not interested in motions of the atmosphere and their relevance to weather prediction. Instead, we had something to say about the general problem of hydrodynamic turbulence. Our claim was that turbulent flow was *not* described by a superposition of many modes (as Landau and Hopf proposed) but by *strange attractors*.

What is an attractor? It is the set on which the point P, representing the system of interest, is moving at large times (i.e., after so-called *transients* have died out). For this definition to make sense it is important that the external forces acting on the system be time independent (otherwise we could get the point P to move in any way we like). It is also important that we consider dissipative systems (viscous fluids dissipate energy by self-friction). Dissipation is the reason why transients die out. Dissipation is the reason why, in the infinite-dimensional space representing the system, only a small set (the attractor) is really interesting.

The fixed point and the periodic loop of Figure 10.4 are attractors, and there is nothing strange about them. The *quasi-periodic* attractor representing a finite number of modes is also not strange (mathematically, it is a torus).[4] But the Lorenz attractor is strange, as are many attractors introduced by Smale (these are more difficult to picture). The strangeness comes from the following features, which are not mathematically equivalent but usually occur together in practice.

First, strange attractors look strange: they are not smooth curves or surfaces but have "non-integer dimension"—or, as Benoit Mandelbrot puts it, they are *fractal* objects.[5] Next, and more importantly, the motion on a strange attractor has sensitive dependence on initial condition. Finally, while strange attractors have only finite dimension, the time-frequency analysis reveals a continuum of frequencies.

The last point deserves further explanation. An attractor representing the flow of a viscous fluid is part of an infinite-dimensional space, but has itself only finite dimension, and is thus well represented by projection in a finite-dimensional space. According to the modes paradigm a finite-dimensional space can describe only a finite number of modes. (Mathematically: a finite-dimensional space can contain only a finite-dimensional torus.) Yet frequency analysis reveals a continuum of frequencies, which one would interpret as a continuum of modes. Is such a thing possible? Can it have anything to do with turbulence?

CHAPTER 11

Chaos: A New Paradigm

Scientists write scientific papers, but they also advertise their ideas and results by giving scientific talks, often called "seminars." A dozen colleagues or more (or less) are assembled and sit for about an hour, listening and looking at equations and diagrams. Some take notes, or pretend to do so but actually work on their own problems. Some appear to doze off, but suddenly wake up with a sharp question. Many seminar talks are hopelessly obscure, because the speaker realizes after half an hour that there is something essential that he forgot to say at the beginning, or because she gets totally mixed up in her calculations, or because he expresses himself in Balkan or Asiatic English of a kind understood by no one else. Yet, seminars are very much at the heart of scientific life. Some are brilliant and illuminating, some are highly polished but insipid, and others, which would appear disastrous to an outsider, are in fact very interesting.

After Takens and I had finished writing our paper on turbulence, I gave a number of talks on this and later work at American universities and research institutions. (I was visiting the Institute for Advanced Study in Princeton during the academic year 1970–1971.) The reception was mixed, but on the whole rather cold. After a seminar that he had invited me to give, I remember the physicist C. N. Yang joking about my "controversial ideas on turbulence"—a fair description of the situation at the time.

What was the cause of the physicists' uneasiness? Well, when a fluid is progressively excited by increasing the applied external forces, the accepted theory predicts a gradual increase in the number of independent frequencies present in the fluid. The prediction of a strange attractor has quite a different effect: a continuum of frequencies should appear. In fact the difference can be tested by making a frequency analysis of some signal produced by a moderately excited fluid. A numerical study was done by Paul Martin at Harvard. Then an experiment was set up by Jerry Gollub and Harry Swinney at City College, New York.[1] The results in both cases favored the Ruelle-Takens rather than the Landau-Hopf picture of the onset of turbulence.

That was the turning point. Not that everybody recognized it at the time, but after the Gollub and Swinney experiment, the controversial ideas progressively became interesting ideas, then well-known ideas. First a few, then many physicists and mathematicians started working on strange attractors and sensitive dependence on initial condition. The importance of the ideas of Edward Lorenz was recognized. A new paradigm arose, and it received a name—*chaos*—from Jim Yorke, an applied mathematician working at the University of Maryland.[2] What we now call chaos is a time evolution with sensitive dependence on initial condition. The motion on a strange attractor is thus chaotic. One also speaks of *deterministic noise* when the irregular oscillations that are observed appear noisy, but the mechanism that produces them is deterministic.

Because of its particular beauty and significance, one result stands out in the theory of chaos—Feigenbaum's period-doubling cascade. Without going into the technical details, I shall try to give an idea of Mitchell Feigenbaum's discovery. When one changes the forces acting on a physical dynamical system, one often sees period doubling, as illustrated in Figure 11.1. A periodic orbit is replaced by another one close to it, but one in

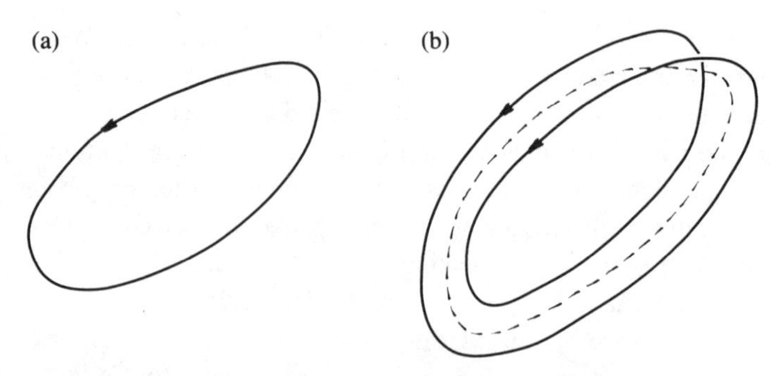

Figure 11.1. Period doubling: (a) projection of a periodic orbit; (b) this orbit is replaced by another one which is approximately twice as long.

which you have to make two turns before coming back exactly at the point of departure. The time it takes to come back—called the period—has therefore approximately doubled. The period doubling is observed in certain convection experiments: a fluid heated from below undergoes some periodic motion; changing the heat setting produces another type of periodic motion with a period twice as long. Period doubling has also been observed in a periodically dripping tap: as the tap is opened more the period doubles (under certain conditions). There are many more examples.

Interestingly, the period doubling can happen again and again, giving a period that is 4 times as long, or 8, 16, 32, 64, . . . times as long. This period-doubling cascade is visualized in Figure 11.2. The horizontal axis measures the forces applied to the system, and the places where the successive period doublings occur are A_1, A_2, A_3, . . . ; they accumulate at a point marked A_∞. Let us now look at the intervals A_1A_2, A_2A_3, A_3A_4, A_4A_5, and so on. They have the property that successive ratios are nearly constant:

$$\frac{A_1A_2}{A_2A_3} \approx \frac{A_2A_3}{A_3A_4} \approx \frac{A_3A_4}{A_4A_5} \approx \ldots$$

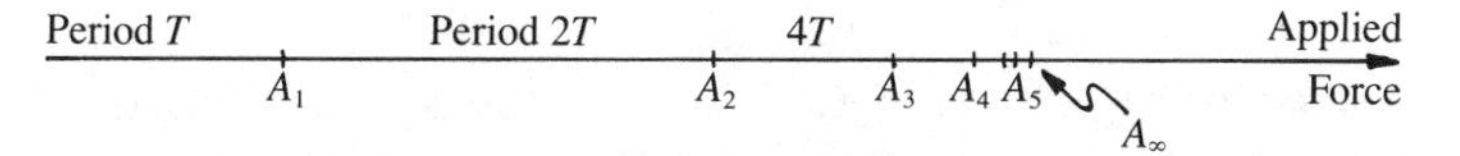

Figure 11.2. The period-doubling cascade. When the forces applied to the system are varied, period doublings occur at the values marked A_1, $A_2, A_3, \ldots$, accumulating at A_∞. Note that for reasons of visibility, the ratio 4.66920 . . . has been replaced in this figure by a smaller value.

More precisely, the following remarkable formula holds:

$$\lim_{n\to\infty} \frac{A_n A_{n+1}}{A_{n+1} A_{n+2}} = 4.66920\ldots$$

After Mitchell Feigenbaum, then a young physicist at Los Alamos, discovered this formula numerically (he was playing day and night with his computer), he proceeded to prove it. For this he followed the ideas of the physicist Kenneth Wilson (then at Cornell) on the *Renormalization Group*. He remarked that the successive period doublings are basically the same phenomenon when properly rescaled (i.e., with suitable changes of the units used for the various parameters of the problem). Working out the necessary rescalings is no easy matter, and Feigenbaum did not give a complete mathematical treatment of the question. This was later supplied by Oscar Lanford (then at Berkeley) along Feigenbaum's ideas. Interestingly, Lanford gave a computer-assisted proof. What it means is that the proof involves some extremely long numerical verifications that would be impractical to do by hand; these are done—quite rigorously—by the computer.

One great interest of the period-doubling cascade is that when you see it in an experiment, you cannot mistake it for something else. Also, beyond the cascade (to the right of A_∞ in Figure 11.2), it is known that there is chaos. Therefore, the observation of the Feigenbaum cascade in hydrodynamics was a particularly convincing proof that modes had to give way to chaos.

I forgot to say one thing. When Mitchell Feigenbaum submitted his paper on the period-doubling cascade for publication in a scientific journal, it was rejected. But then he found a more enlightened editor who accepted it for another journal.[3]

Let us return to strange attractors in turbulence, and remark that we did not use anything specific to hydrodynamics in our discussion, only the fact that a viscous fluid is a dissipative system. We may thus expect to see strange attractors and chaos (or deterministic noise) in all kinds of dissipative dynamical systems. And indeed there are now countless experiments proving this.

But let me go back a bit in time to my own involvement in the chaotic experience. I knew that some chemical reactions proceed in time in an oscillatory way, and that a paper by Kendall Pye and Britton Chance described such oscillations in chemical systems of biological origin.[4] I therefore went, early in 1971, to Philadelphia, where I met with Professor Chance and a group of his collaborators and explained how they might expect to see nonperiodic ''turbulent'' oscillations as well as periodic ones. Unfortunately, the ''mathematical expert'' of the group gave a negative opinion, and Chance dismissed the idea. When I later discussed the matter with Pye, he showed more sympathy, but explained that if he monitored a reaction and got a ''turbulent'' rather than periodic recording, he would declare that the experiment was a failure, tear the recording, and throw it in the wastepaper basket. In retrospect, this story shows what the scientific impact of ''chaos'' has been. When a turbulent or chaotic recording is now obtained, it is recognized for what it is, and studied carefully.

I wrote down my ideas on chemical reactions in a little paper, which I submitted to a scientific journal for publication. It was rejected, but later accepted by another journal.[5] Chaotic chemical reactions were later observed, and led in fact to the first explicit reconstruction of an experimental strange attractor by a group of chemists in Bordeaux.[6]

A few years after the beginnings that I have described, chaos became fashionable, and international conferences were organized on the subject. Then chaos was promoted to the dignity of *Nonlinear Science*, and various institutes were created to study it. New scientific journals appeared, entirely dedicated to Nonlinear Science. The success story of chaos took on the dimensions of a media event, and all scientists in the field should now be jumping up and down with enthusiasm. Some are, and some are not. Let me try to explain why.

Fashions currently play an essential role in the sociology—and in the funding—of physics and other sciences (mathematics being relatively spared). A specialized subject (such as chaos, string theory, or high-temperature superconductors) comes into fashion for a few years, and then is dumped. In the meantime the field has been invaded by swarms of people who are attracted by success, rather than by the ideas involved. And this changes the intellectual atmosphere for the worse.

I shall give only a small personal example of this change. After the publication of my note on chemical oscillations referred to above, a colleague told me, "It is quite a successful paper—I tried to look it up in the university library, and it had been cut out of the journal with a razor blade." I didn't think too much about this, until I received a letter from another university library concerning another paper of mine[7] that had been mutilated by cutting away the first page. (In that case, therefore, it was not a matter of obtaining a cheap copy of a paper, but of making it unusable by others.)

This kind of vandalism remains exceptional. It is characteristic however of a new situation in which the main problem is no longer to convince other scientists that your controversial ideas represent physical reality, but to get ahead of the competition using whatever means are available.

In the case at hand, the mathematical theory of differentiable dynamical systems has benefited from the influx of "chaotic" ideas and, on the whole, has not suffered from the current evolution (the technical difficulty of mathematics makes cheating hard). The physics of chaos, however, in spite of frequent triumphant announcements of "novel" breakthroughs, has had a declining output of interesting discoveries. Hopefully, when the craze is over, a sober appraisal of the difficulties of the subject will result in a new wave of high-quality results.

CHAPTER 12

Chaos: Consequences

I mentioned in the last chapter the low quality of much recent work on chaos. This has unfortunately made the subject disreputable to a number of scientists, including mathematicians who contributed decisively to early research in this area. If however we discard unwarranted claims, and ignore the mass of unusable computations, we find that chaos has given us some remarkably interesting results and insights. I shall now discuss a few examples of applications of chaos and try to give a feeling of what the new ideas are good for.

Remember first that mathematicians have known about sensitive dependence on initial condition since Hadamard's studies at the end of the nineteenth century (and this knowledge has never been forgotten). The computer pictures of new—unexpected—strange attractors have, however, perplexed the experts and given them something to chew on. I wish I could expand on this fascinating topic, but the questions involved are really too technical to be discussed here. I shall similarly omit the discussion of a number of interesting technical topics in physics and chemistry.

Let us return then to the problem of turbulence in fluid motions. Hydrodynamicists would like to have a theory of *fully developed* turbulence: they dream of a very large box full of turbulent fluid. They also dream that you see the same thing if you look at a cubic meter of the fluid, or at a cubic centimeter! More precisely, if you change the scale of length you should see the same thing up to a change of the scale of time. Here again (as in the study of the Feigenbaum cascade) we meet the idea of *scaling*, which pervades modern physics. Does real turbulence satisfy scale invariance? We do not know. There is a good approximate theory of turbulence, the Kolmogorov theory, and it is scale invariant. But this theory cannot be quite correct, because it assumes that turbulence is homogeneous. In reality, a turbulent fluid always shows clumps of intense turbulence in a relatively quiescent background (this is true at all scales!). And hydrodynamicists keep looking for the correct theory to describe this clumpiness.

Strange attractors and chaos have clarified the problem of the onset of turbulence, but not that of fully developed turbulence. What strange attractors have brought is the realization that any theory of turbulence must embody sensitive dependence on initial condition. For instance, in the Kolmogorov theory we no longer have to look for the *period of modes*, but for the *characteristic times* describing how two different histories of the system separate from each other when the initial conditions are almost the same.

Following the early ideas of Edward Lorenz, meteorology has benefited considerably from the notion of sensitive dependence on initial condition. Indeed, according to Lorenz, the fluttering of the wings of a butterfly will change completely the state of the atmosphere after some time (this has been called the butterfly effect).

Now that we have satellite pictures that show the clouds, it is relatively easy (knowing the direction of the wind) to predict the weather a day or two in advance. To go beyond that, meteorologists have developed models of *general circulation of the atmosphere*. The idea is to cover the earth with a grid, identify a certain number of meteorological parameters at each grid point (barometric pressure, temperature, etc.), and then simulate the time evolution of these data on a computer. The initial data (i.e., the values of the meteorological parameters at some initial time) are gathered from satellite, air, and ground observations. The computer then uses these data, the known positions of mountain ranges, and a lot of other information to obtain the values of the meteorological parameters at a later time, and these predictions are then confronted with reality. The conclusion is that it takes about a week for the errors to become unacceptable. Could it be that this is due to sensitive dependence on the initial data? Well, if we try again with slightly different initial values, we find that the two computed time evolutions diverge at about the same rate from each other as they diverge from the time evolution realized by nature. To be honest, the deviation of nature versus computed is a bit faster than that of computed versus computed. There is thus some room for improvement (of the computer program, of the density of the grid used, and of the accuracy of the initial measurements). But we already know that we shall not be able to predict the weather accurately for more than one or two weeks in advance. In the course of their studies, meteorologists have found some situations (called *blocking*) in which the future weather can be predicted more accurately than usual. The control of predictability thus obtained is no mean achievement, conceptually and practically.

Perhaps you start worrying now that some little devil might take advantage of sensitive dependence on initial condition and, by some imperceptible manipulation, upset the carefully planned course of your life. I shall now estimate how much time that would require. The estimates that I shall present are necessarily somewhat tentative, but discussions with colleagues indicate that they are probably not far wrong.

Gravitation, which attracts you to the earth, and the earth to the sun, also acts between the molecules of the air we breathe and every other particle in the world. Our little devil, following a suggestion by the British physicist Michael Berry, will suspend for a moment the attractive effect on our air molecules of a single electron placed somewhere at the limit of the known universe. You will of course not notice anything. But the tiny deflection of air molecules is a change of initial condition. Let us idealize the air molecules as elastic balls and, concentrating on one of them, ask after how many collisions it will miss another molecule that it would have hit if the gravitational effect of the remote electron had been acting. Michael Berry (following an earlier calculation by the French mathematician Emile Borel) has computed that it would take just about 50 collisions![1] After a tiny fraction of a second the collisions of the air molecules will thus have become quite different, but the difference is not visible to you. Not yet.

Suppose that the air that we consider is in turbulent motion (all you need is a little bit of wind); then the sensitive dependence on initial condition present in turbulence will act on microscopic fluctuations of the sort created by the little devil (so-called thermal fluctuations) and magnify them. The net result is that after about a minute, suspending the gravitational effect of an electron at the confines of our universe has produced a macroscopic effect: the fine structure of turbulence (on a millimeter scale) is no longer quite the same. You still don't notice anything, however. Not yet.

But a change in the small-scale structure of turbulence will, in due time, produce a change in the large-scale structure. There are mechanisms for that, and one can estimate the time that they take using the Kolmogorov theory. (As I said, this theory cannot be quite correct, but it will give at least a reasonable order of magnitude.) Suppose that we are in a turbulent part of the atmosphere (a storm would be ideal). We can then expect that in a few hours or a day the imperceptible manipulation of the little devil has resulted in a change of the atmospheric turbulence on a scale of kilometers. This is now quite visible: the clouds have a different shape and the gusts of wind follow a different pattern. But perhaps you will say that this does not really alter the carefully planned course of your life. Not yet.

From the point of view of the general circulation of the atmosphere, what the little devil has achieved is still a rather insignificant change of initial condition. But we know that after a couple of weeks the change will have taken on global proportions.[2]

Suppose then that you have arranged a weekend picnic with your sweetheart (or your boss, I don't care). Just as you have spread your tablecloth on the grass, a really vicious hailstorm begins, arranged by the little devil through her careful manipulation of initial conditions (yes, this little devil is a girl). Are you satisfied now that the carefully planned course of your life may be altered? Actually, the little devil wanted to crash a plane in which you would be flying, but I talked her out of that, out of consideration for your fellow passengers.

Let us return to applications of chaos to the natural sciences. You know that the earth has a magnetic field that acts on compass needles. Once in a while this magnetic field changes its polarity; there are thus periods when the magnetic North Pole is near the geographic South Pole, and vice versa. The reversals of the earth's magnetic field take place erratically at intervals on the order of a million years. (We know about these reversals because they are recorded in the magnetization of certain eruptive rocks that can be dated.) Geophysicists agree that the motion of matter by convection inside the earth sustains electric currents and thus produces the observed magnetic field by a *dynamo mechanism* similar to that of electric generators. But the suggestion is that this strange dynamo has a chaotic time evolution, producing the erratic magnetic polarity reversals that have been documented. Unfortunately, we do not have a good theory that would make this chaotic interpretation quite conclusive.

A rather beautiful and convincing application of chaos is due to the astronomer Jack Wisdom and concerns *gaps* in the asteroid belt between Mars and Jupiter. The belt is composed of many small celestial bodies turning around the sun. But at certain distances from the sun there are no asteroids, there are gaps, and these gaps have baffled students of celestial mechanics for a long time. Most theories that would predict a gap at the right place on the basis of some kind of *resonance* also would predict other gaps where none are seen. The explanation, based on careful computer studies, seems to be as follows. Asteroids in resonant regions have a chaotically varying orbit shape. If this brings the asteroid into the region where the planet Mars circles the sun, a collision occurs and the asteroid is no longer seen. In this way some resonant regions are depleted and replaced by gaps, but others are not, and this can be decided only on the basis of detailed computer calculations.[3]

We now turn briefly to biology. Here is a domain in which we see all kinds of oscillations: chemical oscillations, as in the experiments of Pye and Chance mentioned earlier, circadian rhythms (the alternation of daily activity and rest periods), beats of the heart, waves in electroencephalograms, and so on. The current interest in dynamical systems has stimulated many studies, but the precision attainable in biological experiments is much less than that reached in physics or chemistry, and the interpretation is therefore less certain. If chaos occurs, could it be useful? or just a pathological symptom? Both ideas have been put forward in the case of heartbeats. It is clearly a good idea to study biological systems as dynamical systems, and some excellent efforts are being made in this direction. But there is a lot of inconclusive work, and it seems that we just have to wait for solid results on biological chaos to emerge.

Let me conclude the chapter with some general considerations that will explain the difficulty of studying chaos in biology, ecology, economics, and the social sciences. Quantitative studies of chaos in a system require a quantitative understanding of the dynamics of the system. This understanding is often based on a good knowledge of the basic equations of evolution, which can be integrated with precision on a computer. This is the situation in solar astronomy, hydrodynamics, and even meteorology. In some other cases, like that of oscillating chemical reactions, we do not know the basic equations of motion, but we can monitor the system as a function of time, and obtain long time series of excellent precision. From these time series we can reconstruct the dynamics, if it is simple enough (this is the case for oscillating chemical reactions, but not for meteorology). In biology and in the "soft" sciences, we do not know the basic equations of motion (models that agree qualitatively with the data are not good enough). It is hard to obtain long time series with good precision, and the dynamics is usually not simple. Furthermore, in many cases (ecology, economics, social sciences) the basic equations of evolution, whatever they are, slowly change with time (the system "learns"). For such systems, then, the impact of chaos remains for the time being at the level of scientific philosophy rather than quantitative science.[4] But progress is possible: remember that Poincaré's considerations on predictability in meteorology were just scientific philosophy, and this domain is now quantitative science.

CHAPTER 13

Economics

Chaos, as we have seen, is quite a pervasive feature of natural phenomena. We should thus like to witness its role, at least qualitatively, in economics, sociology, and the history of mankind. Such disciplines indeed offer problems of greater significance to us than gaps in asteroid belts, and even weather predictions. But the analysis of these problems will necessarily be somewhat fuzzy and qualitative. To prepare for this analysis we now review some questions of principle.

First of all, returning to the manipulations by the little devil in the last chapter, you might worry that it is quite impossible to suspend the gravitational action between particles, even for a fraction of a second, and even if they are very far apart. You might also worry that the world we live in is perhaps the only possible world, and that to make any change to it is sacrilegious and inconceivable—it simply makes no sense at all. These worries will disappear if you remember that what we discuss is just an *idealization* of our world. In this idealized description, an absurdly small change leads to important effects after a couple of weeks. Even if you maintain that this says nothing of interest about the "real world," it certainly says something about the intellectual control we have on how it evolves.

In what systems does chaos arise? Suppose you have cooked up an idealized time evolution for a system of your choice. How do you know if it has sensitive dependence on initial condition? If your idealized description is sufficiently explicit to be run on a computer you should by all means run it on a computer, and see if it is chaotic. Apart from that, there are only very vague criteria for the presence of chaos. To describe these criteria, we return for a moment to the *modes* picture discussed earlier. If we have several modes, oscillating independently, the motion is, as we saw, not chaotic. Suppose now that we put a coupling, or interaction, between the different modes. This means that the evolution of each mode, or oscillator, at a certain moment is determined not just by the state of this oscillator at that moment, but by the states of the other oscillators as well. When do we have chaos then? Well, for sensitive dependence on initial condition to occur, *at least three oscillators are necessary*. In addition, *the more oscillators there are, and the more coupling there is between them, the more likely you are to see chaos*.

In general, for the sort of dynamical systems that we consider (continuous-time systems), chaotic time evolution can take place only in a space of at least three dimensions. This is a theorem. Furthermore, introducing interactions between independent systems makes the occurrence of chaos more likely, especially when the coupling is strong (it should not be very strong). This is definitely a vague statement, but quite useful in practice.

Here is another point that the reader should carefully ponder. Although a system may exhibit sensitive dependence on initial condition, this does not mean that everything is unpredictable about it. In fact, finding what is predictable in a background of chaos is a deep and important problem. (Which means that, regrettably, it is unsolved.) In dealing with this deep and important problem, and for want of a better approach, we shall use common sense. Note in particular that living organisms have a remarkable ability to adjust to changes in the environment by regulation mechanisms. One can thus make better predictions about them than the presence of chaos in their environment would suggest. I can, for example, predict that your body temperature is around 37° C, not much below, not much above, or else you wouldn't be reading this book.

A last general remark: the standard theory of chaos deals with time evolutions that come back again and again close to where they were earlier. Systems that exhibit this "eternal return" are in general only moderately complex. The historical evolution of very complex systems, by contrast, is typically one way: history does not repeat itself. For these very complex systems with one-way evolution it is usually clear that sensitive dependence on initial condition is present. The question is then whether it is restricted by regulation mechanisms, or whether it leads to long-term important consequences.

Let us now boldly turn to economics and ask if we can isolate interesting time evolutions: moderately complex and perhaps chaotic. It will be enlightening to examine a scenario of economic development according to the ideas of dynamical systems, and then discuss our findings critically. Our scenario tries to represent the economy of a community at various stages of technological development in parallel with a dissipative physical system subjected to various levels of external forces. For instance, the dissipative system might be a layer of viscous fluid heated from below, and the level of external force would be the degree of heating. We expect of course only a qualitative similarity between the economic system and the physical system.

At low levels of technological development, the economy should have a steady state corresponding to the steady state of a fluid layer weakly heated from below. (A steady state is independent of time—i.e., it is a rather dull state from the point of view of dynamics.) At higher levels of technological development (or heating) we expect that periodic oscillations may occur. In fact, economic cycles, also called business cycles, have been observed and are roughly periodic. At still higher technological levels, we might see a superposition of two or more different periodicities, and economic analysts have indeed seen such things. Finally, at sufficiently high levels of technological development we may have a turbulent economy, with irregular variations and sensitive dependence on initial condition. One may argue that we now live in such an economy.

Pretty convincing, isn't it? Qualitatively, yes. But if we attempt a quantitative analysis we immediately notice that cycles and other fluctuations of the economy take place in a general background of *growth*: there is a one-way historical evolution that we cannot ignore. Business cycles also have their historical features: each one is different; they are not simply monotonous repetitions of the same dynamical phenomenon. If one tries to give a dynamical interpretation of economic phenomena, the ideas of John M. Keynes and his followers come to mind. Most economists would, however, agree now that these interesting ideas have no great predictive value. In other words, economy (specifically, macroeconomy) cannot be analyzed convincingly as a moderately complex dynamical system, even though it has some features of such systems.

I think, nevertheless, that our scenario is not totally wrong, and that it has more than just metaphorical value. The point is that we did not use subtle properties of dynamical systems, but rather robust basic facts. One basic fact is that a complex system (i.e., a system composed of several strongly interacting subsystems) is more likely to have a complicated time evolution than a simple system. This should apply, among other things, to economic systems, and technological development is a way to express complexity. Another basic fact is that the simplest type of time evolution is a steady state: there is no time dependence; the system stays the same. If we assume "eternal return," the next simplest type of time evolution consists of periodic oscillations. Then comes the superposition of two or more oscillations (or modes), then chaos. After removing the background of general growth, one can hope that these remarks apply to economic systems. Our scenario, even if it has little quantitative value, may thus be qualitatively reasonable. We shall now examine one of its consequences.

A standard piece of economics wisdom is that suppressing economic barriers and establishing a free market makes everybody better off. Suppose that country *A* and country *B* both produce toothbrushes and toothpaste for local use. Suppose also that the climate of country *A* allows toothbrushes to be grown and harvested more profitably than in country *B*, but that country *B* has rich mines of excellent toothpaste. Then, if a free market is established, country *A* will produce cheap toothbrushes, and country *B* cheap toothpaste, which they will sell to each other for everybody's benefit. More generally, the economists show (under certain assumptions) that a free market economy will provide the producers of various commodities with an equilibrium that will somehow optimize their well-being. But, as we have seen, the complicated system obtained by coupling together various local economies is not unlikely to have a complicated, chaotic time evolution rather than settling down to a convenient equilibrium. (Technically, the economists allow an equilibrium to be a time-dependent state, but not to have an unpredictable future.) Coming back to countries *A* and *B*, we see that linking their economies together, and with those of countries *C*, *D*, etc., may produce wild economic oscillations that will damage the toothbrush and toothpaste industry. And thus be responsible for countless cavities. Among many other things, therefore, chaos also contributes to the headache of economists.

Let me state things somewhat more brutally. Textbooks of economics are largely concerned with equilibrium situations between economic agents with perfect foresight. The textbooks may give you the impression that the role of the legislators and government officials is to find and implement an equilibrium that is particularly favorable for the community. The examples of chaos in physics teach us, however, that certain dynamical situations do not produce equilibrium but rather a chaotic, unpredictable time evolution. Legislators and government officials are thus faced with the possibility that their decisions, intended to produce a better equilibrium, will in fact lead to wild and unpredictable fluctuations, with possibly quite disastrous effects. The complexity of today's economics encourages such chaotic behavior, and our theoretical understanding in this domain remains very limited.

There is little doubt that economics and finance give us examples of chaos and unpredictable behavior (in a technical sense). But it is difficult to say more, because we do not have here the kind of carefully controlled systems with which physicists like to experiment. Outside events, which the economists call *shocks*, cannot be neglected. Earnest efforts have been made to analyze financial data (which are known with much better precision than economic data) in the hope of isolating a moderately complicated dynamical system. Such hopes, in my opinion, have failed. We are left therefore with the tantalizing situation that we see time evolutions similar in some sense to those of chaotic physical systems, but sufficiently different that we cannot analyze them at this time.[1]

CHAPTER 14

Historical Evolutions

The ideas of chaos apply most naturally to time evolutions with "eternal return." These are time evolutions of systems that come back again and again to near the same situations. In other words, if the system is in a certain state at a certain time, it will return arbitrarily near the same state at a later time.

Eternal return is something you will see in moderately complicated systems, not in very complicated systems. Let me suggest an experiment to prove this point. Take a flea and put it on a particular square of a checkerboard, with a fence to prevent the flea from escaping. Your flea will actively jump around and after a while visit again the square from which it started. This was the case for a moderately complicated system. Now take a hundred fleas and provide them with name or number tags. Put one flea on each square of your checkerboard and watch. How long does it take until all fleas come back simultaneously to the squares from which they started? Intuition (and computations) indicate that it will take such a long time that you will never see this happen. Nor will you see all the fleas simultaneously back in the positions they simultaneously occupied at any earlier time: over any reasonable watching period you will not see the same configuration of fleas twice.

If you do not have a hundred fleas at your disposal, you could do a computer simulation of the experiment, making reasonable assumptions on how fleas jump around. Then you could write a technical paper on your findings, with a title like

"A new theory of irreversibility." If you are going to submit it for publication in a physics journal, don't be bashful. Start boldly with "We have discovered a novel mechanism for irreversibility, etc." or something of that sort, and submit it for publication in the *Physical Review*. They will reject it, of course, and send you copies of three referee's reports saying it's rubbish, and explaining why. Don't be put off; rewrite your paper, taking into account the remarks of the referees, and resubmit it, with a moderately indignant letter to the editors, pointing out the contradictions between the reports of the different referees. After your paper has been sent back and forth a couple more times, driving a few referees crazy, it will be published in the *Physical Review*, and if you were not already a physicist, you will have become one.

Let us now return to eternal return. Why did the word "irreversibility" pop up? Well, if you like the idea of eternal return, ordinary life is full of disappointment: cars get smashed not unsmashed, people get older not younger, and in general the world is different now from what it was before. In brief, things behave irreversibly. Part of the explanation is simply this: if a system is sufficiently complicated, the time it takes to return near a state already visited is huge (think of the hundred fleas on the checkerboard). Therefore if you look at the system for a moderate amount of time, eternal return is irrelevant, and you had better choose another idealization.

Suppose, for instance, that you go back to your hundred fleas and put them all initially on the same square of your checkerboard. They will start eagerly jumping around and will soon occupy the whole board. You may thus put forward the theory that fleas tend to occupy uniformly the space put at their disposal. This is a fairly good theory, in spite of eternal return, and in spite of the fact that the fleas are in fact not interested in occupying the checkerboard uniformly. All they want is to jump around.

If we now look at the complicated world around us, at the evolution of life, at the history of mankind, we should not expect to see eternal return. Eternal return will apply to partial aspects of the world, to small subsystems, but not to the global picture. The global picture follows a one-way historical development, for which we do not have at this point a useful mathematical idealization. (We have some interesting ideas, though, which will be discussed later.) Let us now return to the main theme of this book, chance. We shall try to assess how much the historical development of the world might be altered by absurdly small changes as effected by the little devil of a previous chapter. There are several points that should be carefully discussed, and we shall consider them one by one.

As we have seen, our little devil has no difficulty changing the weather and blowing pollen and seeds one way or the other. The fate of individual plants is thus very much determined by chance. What about animals? Well (as I hope you know), the way animals come into being involves fantastic numbers of little spermatozoa, but only one of them after some kind of race eventually unites with the female gamete. I leave you to ponder the details of the problem, but I think you will come to a distressing conclusion. Namely, that it is only because of the manipulations of the little devil that you have been called into existence rather than some little brother or sister of yours with a somewhat different supply of genes.

But even if individuals are different, the global picture may be much the same. We may be able to predict accurately that a certain soil in a certain climate will support an oak forest, although we cannot predict the position of the trees. In brief, there are many mechanisms of biological regulation, evolutionary convergence, and historical necessity, and they try to erase the eccentricities organized by our little devil. How successful are these mechanisms? Do they lead to historical determinism, i.e., to determinism at the level of the history of large groups of individuals?

Perhaps it is better to speak of partial historical determinism, because the effects of some "chance events" as can be organized by our little devil are not erased by subsequent evolution, but rather fixed seemingly forever. Let us take an example. All known living organisms are related to each other and share essentially the same genetic code. To be specific, the genetic information is written as a sequence of symbols (or "bases") that are the elements of a four-letter alphabet, and each group of three consecutive bases is a code (in principle) for a given building block of a protein (i.e., an amino acid). The coding from base triplets to the twenty different amino acids is arbitrary. If life evolved independently on another planet, nobody would expect it to use the same genetic code. The structure of living organisms has changed a lot through evolution, by the process of mutation and selection, but the genetic code is so basic that it has remained essentially the same from bacterium to humans. Presumably, in the first hesitant steps of life, there was an evolution of the genetic code. When at a certain point an efficient system was evolved, it killed off the competition and survived alone.

This was an example of how an arbitrary feature could be selected forever by historical evolution. There are other examples. Technological evolution, in particular, shows many cases in which choices are made rather accidentally, and then have essentially irreversible long-term effects. Brian Arthur[1] has discussed a number of such situations. He remarks, for instance, that early cars were powered either by internal combustion engines or by steam engines, both being comparably successful. Because of an accidental shortage of the water supply for steam-engine cars, these started to lag behind, so that the internal combustion engine benefited from more technological improvements and supplanted the steam engine. It is a bit difficult to prove such a theory, but Brian Arthur's basic point is undoubtedly correct: of two competing technologies, if one gets ahead, it will benefit from more research and development, and probably soon kill the other one. (This sounds like sensitive dependence on initial condition, although mathematically it is something different.) More generally, it is clear that rather arbitrary decisions, like driving on the right rather than left side of the road, are not easily reversed.

Historical determinism must thus be corrected (at least) by the remark that some historically unpredictable events or choices have important long-term consequences. I think that more can in fact be said. I think that *history systematically generates unpredictable events with important long-term consequences*. Remember indeed that momentous decisions are often taken by individual political leaders. In many cases these political figures act quite predictably under the pressures of the moment. But if they are intelligent and act rationally, the theory of games (as we saw in Chapter 6) will often force them to put a random element in their decisions. Of course, not every kind of erratic behavior is rational, but rational behavior is often erratic in some specific way. The decisions that shape history, when they are taken rationally, involve therefore a random, unpredictable element.

This is not to say that the president of the United States could explain to Congress that he made an important decision by flipping a coin. Maybe that is just what he did, and maybe that was the rational thing to do, but he will have to find something else to say, explaining somehow that there was no reasonable alternative to his decision. In the old days, the political and military leaders were less inhibited, and introduced an element of unpredictability in their decisions by consulting oracles. Admittedly, blind faith in oracles is stupid and can easily have catastrophic consequences. But clever use of oracular unpredictability by an intelligent leader may have been a good way to reach optimal probabilistic strategies.

CHAPTER 15

Quanta: Conceptual Framework

We have just spent several chapters discussing sensitive dependence on initial condition and chaos. We have used for our discussion a certain idealization of physical reality, called classical mechanics, which is largely due to Newton. I have mentioned a couple of times that there is a better idealization, namely quantum mechanics, which originated with Max Planck, Albert Einstein, Niels Bohr, Louis de Broglie, Max Born, Werner Heisenberg, Erwin Schrödinger, and others. For certain aspects of reality (chiefly dealing with small systems, like atoms), classical mechanics is inadequate and has to be replaced with quantum mechanics. But for everyday life Newton's mechanics is good enough, and therefore we do not have to revise our discussion of chaos at that level.

The great philosophical interest of the quantum-mechanical description of the world is this: chance plays an essential role in it. I shall try to show how this comes about.

Quantum mechanics, like other physical theories, consists of a mathematical part, and an operational part that tells you how a certain piece of physical reality is described by the mathematics. Both the mathematical and the operational aspects of quantum mechanics are straightforward and involve no logical paradoxes. Furthermore, the agreement between theory and experiment is as good as one can hope for. Nevertheless, the new mechanics has given rise to many controversies, which involve its probabilistic aspect, the relation of its operational concepts with those of classical mechanics, and also something called the collapse of wave packets. These controversies still go on to some extent, and the technical character of the mathematics involved complicates the discussion.

If you have not had a course in quantum mechanics, and even if you have, I recommend that you read Feynman's little book called *QED*.[1] It tells as much as is possible to tell about the conceptual structure of the subject without using technical mathematics. Here I shall be more modest, and give only a skeleton of the theory. This skeleton is not very funny: brace yourself and try to show fortitude while reading the next couple of pages.

Remember that in classical mechanics we had positions and velocities as basic notions, and Newton's laws told us how positions and velocities evolve with time. We also discussed probabilistic theories, in which the basic objects are probabilities, and we could have set up laws dictating how these probabilities evolve with time. Quantum mechanics has basic objects called *amplitudes* (or *probability amplitudes*—we shall see why in a moment). These amplitudes are complex numbers, instead of the more usual real numbers.[2] The mathematical part of quantum theory dictates how the amplitudes evolve with time; the evolution equation is called the Schrödinger equation. This is a rather straightforward but technical piece of mathematics, to which only a note can be dedicated here.[3] Note that the evolution of amplitudes is deterministic. The mathematical part of quantum theory also contains objects called *observables*. Technically, these are *linear operators*, and their abstract character very much impressed the first physicist who used them. Finally, given an observable—call it A—and a set of amplitudes, one can compute a number called the *mean value of* A, which we shall denote by $\langle A \rangle$.[4]

To summarize, quantum mechanics tells us how to compute the time evolution of amplitudes, and then how to use these amplitudes to obtain the mean value $\langle A \rangle$ of an observable A.

How do we connect these mathematical concepts with physical reality? Let us be specific and suppose that you are an experimental particle physicist: you like to accelerate particles to large energies, aim them at a target, and see what comes out. You have surrounded your target with a number of detectors, I, II, III, and so on, that will click when a particle of the right kind hits them at the right time. (The "right kind" means the right charge, the right energy, etc. The "right time" means that, for instance, detector II is activated only if I has clicked, and then only for a definite interval of time.) You decide to call *event* A the situation in which I and II click and III does not click. (Event A is the signature of a particular type of collision that you expect to see in your experiment.)

You now go and consult the Holy Scriptures of quantum mechanics, and these will tell you which observable corresponds to event A. (Events are thus viewed as a special kind of observable.) The Holy Scriptures will also tell you how to compute the amplitudes relevant to your experiment. Then you will be able to estimate $\langle A \rangle$. A fundamental dogma of quantum faith is that $\langle A \rangle$ is the probability that you will see the event A. Specifically, if you repeat your experiment a large number of times, the proportion of cases in which all detectors will click as required is $\langle A \rangle$. This is the connection between the mathematics of quantum theory, and the operationally defined physical reality.

Let me remark in passing that some chapters of the Holy Scriptures of quantum mechanics have not yet been written, or only tentatively. In other words, we do not know for sure all the details of the interactions between particles, and this is why experiments are still being performed.

Later we shall try to develop some physical intuition about

quantum mechanics, but the schematic description just presented will be adequate for discussing basics. Let me repeat what the setting is: A physical process is taking place (a collision between particles, say), which we study by making a certain number of measurements (using detectors, say). The totality of measurements represents an event, and quantum theory allows us to compute its probability. (There is nothing magic about a measurement: if you would like to understand what is going on in a detector, you may surround it with other detectors, make measurements, and apply quantum mechanics to that.) In this way we obtain a description of the world that is profoundly different from the description given by classical mechanics, but completely consistent.

If you want to say that quantum mechanics is deterministic, it is: the Schrödinger equation predicts unambiguously the time evolution of the probability amplitudes. If you want to say that quantum mechanics is probabilistic, you may: the only predictions are about probabilities. (The probabilities are occasionally 0 or 1, and then you have certainty, but this is usually not the case.)

While quantum mechanics is probabilistic, it is not a probabilistic theory in the usual sense discussed in Chapter 3. Specifically, when event "*A*" and event "*B*" are defined in an ordinary probabilistic theory, an event "*A* and *B*" is also defined (with the intuitive meaning that "*A* and *B*" occurs if "*A*" occurs and "*B*" occurs). In quantum mechanics, "*A* and *B*" is usually not defined: there is no entry for "*A* and *B*" in the Holy Scriptures of quantum mechanics. This is very irritating, of course: why can't we just say that "*A* and *B*" occurs if "*A*" occurs and "*B*" occurs? There is a dual answer to this question—mathematical and physical-operational. Physically, what happens is that (in general) you cannot pick detectors to measure "*A*" and "*B*" simultaneously (i.e., check at the same time if "*A*" occurs and "*B*" occurs). You can try to measure first "*A*" then "*B*," or first "*B*" then "*A*," but you get different answers! This is often expressed by saying that the first measurement perturbs the second. This intuitive interpretation is not really wrong, but it is somewhat misleading: it suggests that the event "*A* and *B*" in fact makes sense, but we are too clumsy to measure it. The mathematics of quantum theory is, however, unambiguous: "*A* and *B*" does not usually make sense. This has to do with the fact that the observables *A* and *B* "do not commute," and some more details are given in a note.[5]

All this talk about events is a bit abstract. What can we say about a particle moving on a straight line? According to classical mechanics, all we would want to know about it is its position x and its velocity v. How is it in quantum mechanics? Suppose that your particle is described by certain probability amplitudes. By looking at the events "x is here," "x is there," you can determine the probabilities of finding the particle in various places. (As it happens, the various events concerning x are commuting observables, and you can observe them simultaneously.) Let us summarize your findings by saying that the particle is near x_0, but that there is an uncertainty (or probable error) Δ_x in its position. Similarly, you can summarize the probabilistic description of the velocity of the particle by saying that it is near v_0 with an uncertainty Δ_v. If the particle were described by probability amplitudes such that Δ_x and Δ_v were both zero, then its position and velocity would be perfectly well defined. But this is impossible, because "x" and "v" are not commuting observables, and Werner Heisenberg proved in 1926 that

$$m\,\Delta_x \cdot \Delta_v \geq h/4\pi,$$

where m is the mass of the particle, $\pi = 3.14159\ldots$, and h is a very small quantity called *Planck's constant*. The above inequality is the celebrated Heisenberg uncertainty relation. It brings out vividly the probabilistic character of quantum mechanics.

But, as we have said, quantum mechanics is not an ordinary probabilistic theory. The physicist John Bell showed that the probabilities attached to a simple physical system satisfy some inequalities that are in fact incompatible with an ordinary probabilistic description.[6] Bell's result shows how far quantum mechanics is from usual intuition.

Of cource, there have been valiant efforts (in particular, by the physicist David Bohm) to bring quantum mechanics closer to classical ideas. Such efforts are respectable and necessary. But what has been achieved involves somewhat unnatural constructions, and it remains unconvincing to most physicists. One bit of effort to bring quantum mechanics closer to usual intuition has found its way into the Holy Scriptures . . . and has caused a lot of trouble. This is the sacred Dogma of the Collapse of Wave Packets. This has to do with the successive measurement of two observables A and B, and proposes to say what the probability amplitudes are after the measurement of A, and before that of B.

But the dogma leads to difficulties, and is best left aside. (From the point of view of physics, all that counts is that you be able to assess the probabilities associated with "A and then B.")

In recent times, and with due reverence to the Founding Fathers who wrote the Holy Scriptures, physicists have tended to stay away from the collapse of wave packets. Richard Feynman, for instance, mentions the topic only in a brief footnote in his book *QED*, just to say that he doesn't want to hear about it.[7]

CHAPTER 16

Quanta: Counting States

The conceptual skeleton of quantum mechanics that we examined in the last chapter did not have much physical meat attached to it. Here, in effect, is what we found: quantum mechanics gives rules for computing the probabilities of events. It is thus a probabilistic theory, but not a standard one, because, given events "*A*" and "*B*," the event "*A* and *B*" often does not make sense.

The meat of quantum mechanics is of course in the rules, in their application to specific problems, and in the physical insight thus achieved. This is not the place to go into a technical discussion of quantum mechanics, but it is easy and rewarding to develop a bit of physical intuition. Just keep in mind that when physicists develop an intuitive argument, they back it with hard calculations. Nontechnical expositions of science, since they avoid such hard calculations, are always somewhat mystifying; at the technical level things are less simple, but also less mysterious.

I want to present a little calculation now, using nothing more than high-school mathematics and physics. This calculation is not really indispensable for what follows, but it is well worth doing anyway.

We consider, as in the last chapter, a particle of mass m moving along a straight line, but we now put the particle in a box.

More precisely, we constrain the position x of the particle to be in an interval of length L. We also constrain the velocity v of the particle to be between $-v_{max}$ and v_{max} (the particle has speed at most v_{max} and may go left or right). Drawing a diagram of the position x and the product mv (mass times velocity), we see that the allowed region for the particle is the large rectangle of Figure 16.1. But we can choose a state of the particle such that it is concentrated in a smaller region—the little shaded rectangle with sides "l_x and ml_v. For this state, the position x is known with an uncertainty about $\frac{1}{2}l_x$ and the velocity with an uncertainty about $\frac{1}{2}l_v$. In agreement with the Heisenberg uncertainty relations we thus have to choose l_x and l_v such that $ml_x \cdot l_v \geq h/\pi$. In fact, a more careful study shows that the best one can do is to take

$$ml_v \cdot l_x = h,$$

i.e., the shaded rectangle has area h. The space of the variables x and mv is called *phase space*. We have drawn another little rectangle in phase space, not overlapping with the first, and therefore corresponding to a completely different state of our particle. How many completely different states are there? The number is the area of the large rectangle divided by the area of the small rectangle, i.e.,

$$\text{number of different states} = \frac{2m\, v_{max} \cdot L}{h}.$$

A serious, technical computation would confirm this result.[1] Note that while the number of different states is well defined, these states can be chosen in various ways (the area of the little rectangles is fixed and equal to h, but their shape can be chosen in different manners).

Let us now look at the energy of our particle; I mean the

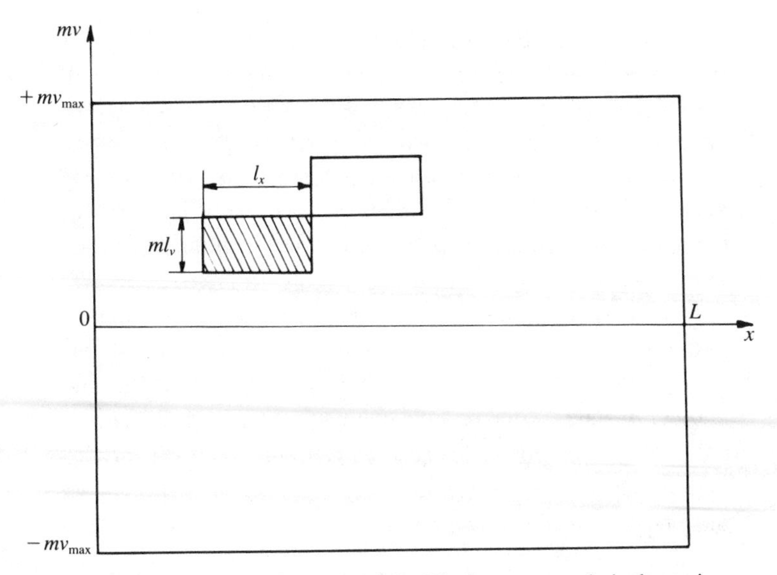

Figure 16.1. Phase space of a particle. The large rectangle is the region accessible to the particle. The small rectangle measures the indeterminacy imposed by quantum uncertainty.

energy due to its velocity, called the kinetic energy. If you hold a driver's license from a country or state with high intellectual standards, you may have learned the kinetic energy formula for your driver's test. Here is the formula:

$$\text{energy} = \frac{1}{2}mv^2.$$

(The kinetic energy is one half the mass times the velocity squared: if you smash a car of mass m at speed v into a wall, this is how much energy is available to wreck the wall and your car and send yourself to the hospital.) Saying that our particle has velocity between $-v_{max}$ and v_{max} means therefore that it has (kinetic) energy at most $E = \frac{1}{2}mv_{max}^2$.

In conclusion, if we constrain a particle to be in a box and to have less than a certain energy, then it has only a finite number of different states. There is a certain arbitrariness in how to choose these states, but a technical study shows that they can be taken to have *precisely defined energies*. This is expressed by saying that the *energy is quantized*: it can take only discrete values. The quantization of energy is a characteristic feature of quantum mechanics, and quite contrary to the intuition of classical mechanics.

Instead of a particle on a line we may consider an honest particle moving in three-dimensional space and put it in an honest box of some volume V. It is then possible to compute the number of states of the particle that have energy less than some value E. (This involves using three Heisenberg uncertainty relations for the three directions of space.) For the hell of it let me give you the formula that one can derive:

$$\text{number of states} = \frac{1}{h^3} \cdot \frac{4}{3}\pi \left(\frac{2E}{m}\right)^{3/2} \cdot V.$$

The seasoned scientist will immediately recognize that this is the accessible volume of phase space, measured in units of h^3. Here the phase space has six dimensions: it gives the position x of the particle, and also the vector mv (mass times velocity).

The notion that you can say something profound about the physical universe by manipulating a few symbols like h or π is suggestive of witchcraft. As a result, a formula like the one above elicits intense revulsion in some people, and immoderate enthusiasm in others. Physicists are of course among the enthusiasts, accepting with dedicated professionalism their roles as modern sorcerers. For the purposes of this book I shall however have to forsake professionalism and proceed mostly without formulas.

But I see that you want to return to the counting of states. You want to put now not one but many particles in a box. You have in mind using molecules of oxygen, nitrogen, helium, or some other gas as your particles, and looking at one liter of the gas in question. At normal temperature and pressure, this is about 2.7×10^{22} molecules, i.e., 27000000000000000000000 (twenty-three digits). Your pocket computer may use the notation 2.7E22 for this number. (Popular science writers like to say: twenty-seven thousand million million millions, but no one else uses this clumsy language.) Anyway, you want to know how many different states there are for a system of 2.7E22 molecules of helium in a one-liter container. You still have to tell me what the total energy of your liter of helium is. A reasonable choice is the total energy corresponding to the motion of helium particles at room temperature. In other words, you want to count the number of quantum states in which a liter of helium at room temperature can be found. (Instead of saying "at room temperature," we should really be saying "with a total energy not greater than that contained in a liter of helium at room temperature." But, as it happens, this doesn't really make any difference for the answer.)

Here is the answer[2]

number of states = 1E50000000000000000000000

Of course 5 followed by 22 zeros is better written as 5E22. But there is already an "E" above; isn't this a mistake? In fact, no. The number of states has a number of digits which is 5E22, and might therefore be written 1E5E22. If you had to write this number in full on a sheet of paper, you would need a very large sheet of paper, and you would be dead before you would be through with your writing task.

Numbers like 1E5E22, so far away from ordinary intuition, again elicit intense revulsion in some people, and immoderate enthusiasm in others. A reasonable attitude is to make the following definition:

entropy = number of digits of the number of states
(= 5E22 in this case).

CHAPTER 17

Entropy

There are several ways to do hard scientific thinking. Some people just sit at their desks and stare at a sheet of paper; others pace around. I personally like to lie flat on my back, with eyes closed. A scientist who is working really hard may look very much like one who is taking a little nap. Hard scientific thinking is a most rewarding experience, but it is also hard work. One has to pursue ideas relentlessly, allow oneself to become obsessed. If an interesting possibility seems to appear, it has to be brought into focus, verified, sometimes retained but mostly rejected. Bold general ideas have to be developed, but details must then be checked, and all too often, disastrous flaws are discovered. The construction must then be rearranged, or large parts of it discarded. And the process goes on day after day, week after week, month after month. Not all those who pose as scientists work hard, of course. Many stopped working a long time ago; others have never started. But for those who really play the game, rather than just fool around and pretend, the game is hard, painful, strenuous, exhausting. And if the fruit of this labor, the result of this exertion, is received with arrogance and disdain, then tragedy may follow. Imagine a man who has found the meaning of a fundamental aspect of Nature. Year after year he has pursued his research in spite of the attacks and the misunderstanding of his contemporaries. He is now getting old, ill, and depressed. This is what happened to the Austrian physicist Ludwig Boltzmann. On September 5, 1906, he committed suicide; he was sixty-two years old.

Boltzmann and the American J. Willard Gibbs were the creators of a new science called statistical mechanics. Their contribution is not less important for twentieth-century physics than the discovery of relativity or quantum mechanics, but it is of a different nature. While relativity and quantum mechanics destroyed existing theories and replaced them with something else, statistical mechanics operated a quiet revolution. It built upon existing physical models, but established new relations and infused new concepts. The conceptual machinery developed by Boltzmann and Gibbs has proved extraordinarily powerful and is now applied to all sorts of situations far beyond the problems of physics originally addressed.

Boltzmann's starting point was the atomic hypothesis: the notion that matter is composed of a huge number of little balls dancing around. In the late nineteenth century, when Boltzmann was active, the atomic structure of matter was still unproved and far from generally accepted. Part of the attacks against Boltzmann were motivated by his belief in atoms. Not only did he believe in their existence, but he proceeded to derive striking consequences from the assumed atomic structure of matter.

Only classical mechanics was available during Boltzmann's time, yet it is convenient to present some of his ideas in quantum language. After all, there is a close relation between classical and quantum mechanics. Both try to describe the same physical reality, and, for instance, a *number of states* in quantum mechanics corresponds to a *volume of phase space* in classical mechanics. I shall thus concentrate on ideas, and not worry too much about anachronisms of detail.

The industrial revolution of the nineteenth century had brought great interest in the steam engine and in the transformation of heat into mechanical work. It was known that you can freely transform mechanical energy into heat (for instance by rubbing two stones), but not the converse. Heat is a form of energy, but its use follows rather strict rules: some processes take place easily, others not at all. For instance, it is easy enough to mix a liter of cold water and a liter of hot water to obtain two liters of lukewarm water. But try to unmix the two liters to get back one liter of cold water and one of hot water! It doesn't work; the mixing of cold and hot water is an irreversible process.

One step was taken toward understanding irreversibility when *entropy* was defined (forget for a minute that we already used this word in the last chapter). A liter of cold water has a certain entropy, and a liter of hot water has a different entropy. These entropies can be computed from experimental data, but we won't bother to specify how. The entropy of two liters of cold water is twice the entropy of one liter, and similarly for hot water.

If you put one liter of cold water and one liter of hot water side by side, the sum of their entropies has a certain value. But if you now mix the two, the entropy of the two liters of lukewarm water obtained has a larger value. By mixing cold and hot water, you have increased the entropy of the universe—irreversibly. Here is the rule, known as the *second law of thermodynamics*:[1] in every physical process, the entropy stays constant or increases, and if it increases the process is irreversible.

This is of course all rather mysterious, and not totally satisfactory. What is the meaning of entropy? Why does it always increase and never decrease? These were the problems that Boltzmann tried to solve.

If you believe in the "atomic hypothesis," the molecules composing a liter of cold water can be in all kinds of different configurations. In fact, the molecules dance around and the configuration changes all the time. In quantum language, we have a system of many particles, which can be in a very large number of different states. But while these states would look different if you could see microscopic details, they all look the same to the naked eye; in fact, they all look like a liter of cold water.

So, when we refer to a liter of cold water, we refer in fact to something quite ambiguous. Boltzmann's discovery is that the entropy is a measure of this ambiguity. Technically, the right definition is that the entropy of a liter of cold water is the number of digits in the number of "microscopic" states corresponding to this liter of cold water. The definition extends of course to hot water, and to many other systems. This is in fact how we defined the entropy of a liter of helium in the last chapter.

But the definition of the last chapter was not physically motivated. Boltzmann's achievement was to relate a natural mathematical concept and a previously mysterious physical quantity. Technically, one would want to say "logarithm" instead of "number of digits," multiply by a constant k (k is in fact called Boltzmann's constant), and perhaps add another constant to the result, but this is not the place to discuss such details.

Let us now put side by side a liter of cold water and a liter of hot water, without mixing them. Each state of the liter of cold water and each state of the liter of hot water give a state of the composite system. Therefore the number of states of the composite system is the product of the numbers of states of the two component liters, and the entropy is the sum of the entropies. This is not too astonishing; the definitions have just been set up this way.

But what happens if we mix the cold and the hot water? Somehow, we obtain lukewarm water, and the details of exactly how this happens still puzzle scientists. What is well established is that the number of states of two liters of lukewarm water is larger than the number of states of one liter of cold water and one liter of hot water. And remember that all the states of lukewarm water look the same to the naked eye: there is no way to recognize those states that come from mixing cold and hot water. The entropy therefore increases as a result of the mixing.

But why should there be irreversibility? The world around us behaves very irreversibly, but how do we prove that it must do so? In science, when you don't see how to prove something, it is often a good idea to try to disprove it, and see what happens. So, let me try to arrange reversibility.

The basic laws of classical mechanics do not contain any irreversibility. Suppose that you watch the motions and collisions of a system of particles for one second, and suppose that you could then suddenly reverse the velocities of all the particles. They would then go backward, and collide again in reverse order, and after another second you would be back to the initial condition (with velocities reversed, but if you like, you could reverse those velocities again). By this argument, if the entropy goes up, it can also go down, and irreversibility is impossible. Was Boltzmann wrong? Or did we miss something?

What we have arranged is to have the time "run backward" by simultaneously reversing the velocities of all the particles in a large system. One can of course argue that this is impossible in practice. But something like this is possible for some systems (spin systems). And of course, it is embarrassing to base a general law of physics, like the fact that the entropy always increases, on a practical impossibility that might be lifted some day.

There is, however, a subtler impossibility about the velocity-reversing experiment just described, and this has to do with sensitive dependence on initial condition. When we apply the laws of classical mechanics to study the motions and collisions of a system of atoms or molecules, we imagine that the system does not interact with the rest of the universe. But this is quite unrealistic. Even the gravitational effect of an electron at the limit of the known universe is important and cannot be neglected. If we reverse velocities after one second, we do not see time running backward. After a small fraction of a second, the electron at the limit of the universe will have changed the course of things, and entropy, instead of decreasing, will continue to increase.

The role of sensitive dependence on initial condition in understanding irreversibility was in fact not appreciated in Boltzmann's time. Again, I allowed a little anachronism in my discussion. Retrospectively, we see that Boltzmann's ideas fit nicely with what we have later learned to be true. But during Boltzmann's time, things were far from clear. Of course he knew that he was right. Others saw that Boltzmann's work was entirely based on the dubious "atomic hypothesis." They saw that he used dubious mathematics to derive an irreversible time evolution from the laws of classical mechanics, which are clearly reversible. They were unconvinced.

CHAPTER 18

Irreversibility

I have stressed that the aim of physics is to give accurate mathematical descriptions of pieces of physical reality, and that one should not worry too much about the "ultimate truth," whatever that is. This may sound rather unambitious, and you may think that studying physics must therefore be a rather dull enterprise. Quite the opposite is true, however, because physical reality is itself far from dull. Discussing physics *in abstracto*, and without reference to the world it tries to explain, is thus misleading and rather useless.

A case in point is that of Boltzmann's ideas. He started from *thermodynamics*, the theory that deals with entropy and irreversibility. Thermodynamics fitted and still fits the experiments very well. The great enterprise of Boltzmann's life was to give an interpretation of thermodynamics in the framework of the "atomic hypothesis" by doing statistical mechanics. If atoms had been forever elusive, and if statistical mechanics had never had more predictive value than in Boltzmann's time, then it would make little sense to a physicist that his ideas were "true." But Boltzmann's vision has come true, because it is now proved that matter is composed of atoms, because Boltzmann's formula for the entropy can be checked experimentally, and because statistical mechanics has acquired enormous predictive value (largely through the efforts of Gibbs and later physicists).

As it happens, Boltzmann's ideas about atoms were far from

the ultimate truth. Atoms are not simply little balls dancing around; they have a rather complicated structure and require quantum mechanics for their description. Boltzmann's preconceptions have served him (and us) well, but they provide only one step in our understanding of Nature. Will there be a last step? Is there an ultimate truth in physics? Hopefully, the answer is positive, and the ultimate physical theory of matter will be discovered (and proved to be correct) in our lifetime. But it should be clear that the importance of Boltzmann's ideas does not depend on the eventual discovery of the ultimate physical theory.

There is something romantic about Boltzmann's life. He killed himself because he was, in some sense, a failure. Yet he is now considered one of the greatest scientists of his time, much greater than those who were his opponents. Obviously, he was right too early. But how does one manage to be right early? I think that part of the answer is prejudice. One needs some preconceived ideas about physics, different from the generally accepted dogma, and one should follow these ideas with some obstinacy. Perhaps these are the same ideas that were proved wrong on earlier occasions, but if you have the right insight, and if you are lucky, these ideas will give you the keys to a new understanding of Nature. Boltzmann's prejudice was definitely mechanistic. Descartes had been pushed earlier by a similar mechanistic prejudice and got nowhere, while Newton, with a different prejudice, founded modern physics. But in Boltzmann's time, the mechanistic prejudice was the right one for understanding thermodynamics, and it worked. Here are some other examples of prejudice: that mathematics is the language of nature (Galileo); that our world is the best of all possible worlds (Leibniz); that the laws of Nature must satisfy aesthetic requirements (Einstein). At any given time some preconceived ideas about science are fashionable, others are not fashionable but may make you famous after your death . . .

I shall now interrupt these considerations of posthumous glory and return to our unfinished discussion of irreversibility. Let us look again at the time evolution of a complicated system of particles, like the atoms of helium in a one-liter container, or the molecules in one liter of water. We shall use classical mechanics to describe our particles, and assume that they form an isolated system: there is no interaction with the outside world, and therefore no energy is received or given out. Boltzmann had the idea that over the course of time, the system would visit all energetically possible configurations. In other words, all configurations of positions and velocities of the particles that have the right total energy would be realized, and you could observe them if you waited long enough. More correctly, the system would come close (again and again) to any energetically possible configuration; this is an example of what we called *eternal return* earlier. The proper mathematical formulation of Boltzmann's idea, known as the *ergodic hypothesis*, is not very easy, and was achieved only after his death. But the physics is clear enough, and well worth understanding.

You should remember that when the quantum physicist speaks of *number of states*, the classical physicist must speak of *volume of phase space*; this is now the relevant notion. In the example of a liter of helium, a point in phase space specifies all the positions and velocities of the helium atoms. We shall restrict our interest to the part of phase space that consists of configurations with a given total energy (because our system does not receive or give out energy). We shall think of the time evolution of our complicated system as described by the motion of a point in phase space. We are now in a position to express the content of the ergodic hypothesis: *in its motion through phase space, the point representing our system spends in each region a fraction of time proportional to the volume of the region.*[1]

If we accept the ergodic hypothesis we may now understand why, when we have two liters of lukewarm water in a bottle, we never see the liquid spontaneously separate into a layer of cold water and a layer of hot water. Indeed, as we saw earlier, the entropy of two liters of lukewarm water is greater than that of a liter of cold and a liter of hot water. Let us suppose that the difference in entropies is 1 percent. This means that if we first count the states of a liter of cold and a liter of hot water, then the states of two liters of lukewarm water, we get two huge numbers that differ in length (number of digits) by 1 percent. The numbers of states, or volumes in phase space, differ therefore by a huge factor. Specifically, the volume in phase space for two liters of lukewarm water is very much greater than the volume for one liter of cold and one liter of hot water. Let us now observe the point representing our system as it moves through phase space. According to the ergodic hypothesis it will spend most of its time in the region corresponding to two liters of lukewarm water. Very little time will be spent in the part of phase space corresponding to a layer of cold water and a layer of hot water; in practice you will never see lukewarm water separate into cold and hot water.

Let me repeat the explanation. You have carefully poured a layer of hot water over a layer of cold water. In this way you have arranged that your system is in a small special region of its phase space. After a little while, heat will have diffused and you have homogeneous lukewarm water, corresponding to a much bigger region of phase space. If you wait long enough, eternal return will bring your system back to a layer of cold water and a layer of hot water. But how long is long enough? The method of estimating this time is related to counting states as we did in Chapter 16, and the answer is dishearteningly large. Long enough is just too long. Because of the brevity of life, we shall never again see a layer of hot water over a layer of cold water, and in this sense the mixing of the two layers is irreversible. (For the role of sensitive dependence on initial condition, see the note.)[2]

The explanation of irreversibility that we have obtained, following Boltzmann, is at the same time simple and rather subtle. It is a probabilistic explanation. There is no irreversibility of the basic laws of physics, but there is something special about the initial state of the system that we are considering: this initial state is *very improbable*. By this we mean that it corresponds to a relatively small volume in phase space (or a small entropy). The time evolution then leads to a region with relatively large volume (or large entropy), which corresponds to a very probable state of the system. In principle, after a very long time the system will return to the improbable initial state, but we shall not see this happening . . . As a physicist, you will want to make an idealization in which the number of particles in your system tends to infinity and the time of eternal return also tends to infinity. In this limit you have true irreversibility.

I have described the interpretation of irreversibility that is now generally accepted by physicists. There are some dissenting voices, such as that of Ilya Prigogine,[3] but the disagreement is based on philosophical prejudice rather than physical evidence. There is nothing wrong with philosophical prejudice; it is invaluable in making discoveries in physics. But in due time things have to be settled by careful comparison of mathematical theories and physical experiments.

One of the ingredients of our discussion, the reversibility of the basic laws of physics, seems to be a good assumption.[4] But what about the ergodic hypothesis? It would require a mathematical proof, and such a proof is still missing, even for simple models. Physicists, however, do not worry too much about this. It is realized that many important mathematical and physical aspects of our understanding of irreversibility have to be made more precise. The ergodic hypothesis probably has to be weakened. Another way of looking at things may be needed for some systems, such as *spin glasses*. Basically, however, we think we understand what is going on.

This confidence may someday be shaken, but for the moment, it receives support from our good understanding of *equilibrium statistical mechanics*. This branch of physics does not worry about the complex problem of mixing cold and hot water, only about comparing cold water with hot water, and also with ice and water vapor. The predictions of equilibrium statistical mechanics are in very precise agreement with experiment. Clearly, this is a domain of physics in which we know what we are doing. Equilibrium statistical mechanics is a rather technical subject, and one that is conceptually very rich. Its powerful ideas have been transferred to mathematics and to other parts of physics, where they play a major role. I see equilibrium statistical mechanics somehow as science at its best, and I shall therefore try to give you a glimpse of the subject in the next chapter.

CHAPTER 19

Equilibrium Statistical Mechanics

You visit an art museum and walk through the early twentieth-century French painting section. Here is a sumptuous Renoir, there an unmistakable Modigliani, there some flowers by van Gogh and fruits by Cézanne. Further on you have a glimpse of a Picasso, or perhaps it is a Braque. You have not seen these paintings before, but you usually have no doubts as to who the artist was. Van Gogh, in the last years of his life, painted an amazing number of works, all stunningly beautiful, and immediately distinguishable from paintings by Gauguin, for instance. How do you tell the difference? Well, the paint is not applied in the same manner, and the subjects treated are different, but there is something else, more difficult to state explicitly yet immediately recognized, that has to do with the texture of shapes and the balance of colors.

Similarly, if you turn on the radio you will immediately know if you are hearing classical music or the Beatles. And if you have the slightest interest in classical music you can distinguish Bach from sixteenth-century music, Beethoven from Bach, and Bartok from Beethoven. You may not have heard the pieces before, but there is something unique about the arrangement of sounds that allows almost instant recognition. One can try to capture this ''something unique'' by statistical studies.[1] In particular, one can study the intervals between successive notes. Small musical intervals are particularly common, but they are most common in older music. Recent music uses all kinds of intervals more randomly. Evaluating the frequency of intervals between successive notes in a piece of music, we may thus decide whether it is by Buxtehude or Mozart or Schönberg. Of course, we shall reach the same conclusion even more accurately and much more quickly by listening to a few bars. But this is in fact using the same method: the human ear-brain system is a marvelous device for extracting the sort of statistical information that allows us to say, This is music by Monteverdi, or by Brahms, or by Debussy.

My contention is thus that we base our identification of a painter or a composer on statistical evidence. But you may think that this is absurd: how can we be sure of an identification if we base it on probabilities? The answer is that we can be almost sure. Just as we are often almost sure about the sex of a person whom we meet on the street: men are usually taller, have shorter and darker hair, larger feet, and so on. Any single characteristic is rather unreliable, but you evaluate many characteristics in a split second, and that often leaves no reasonable doubt.

A question remains, however: Why is it that a given artist repeatedly produces works with the same cluster of probabilistic features, which characterizes that particular artist? Or to take another example: Why is it that your handwriting is so unique, so hard for others to imitate or for you to disguise? We don't know the answers to these questions because we don't know in detail how the brain functions. But we understand something very similar, a basic fact that is in some sense the cornerstone of equilibrium statistical mechanics.

Here is the basic fact: *If one imposes a simple global condition on a complicated system, then the configurations satisfying this condition usually have a cluster of probabilistic features that uniquely characterizes these configurations*. Read the above sentence again: it is deliberately vague and metaphysical, so that it could be applied to painting or music. Authorship by a certain artist is then the "simple global condition," and the "cluster of probabilistic features" allows us to identify the artist. But now we want to discuss the situation of equilibrium statistical mechanics. Here, the complicated system will typically consist of a large number of particles in a box (a liter of helium is our standard example). And the simple global condition will be that the total energy of the system has at most some value E. We restrict the *macroscopic* state of the system, and this, allegedly, will determine its *microscopic* probabilistic structure.

Let me again yield to the urge to write an equation. Here is the expression for the energy of a system of particles in terms of the velocities v_i of the particles, and their positions x_i:

$$\text{energy} = \sum_i \frac{1}{2} m v_i^2 + \sum_{i<j} V(x_j - x_i).$$

As we saw earlier, $\frac{1}{2} m v_i^2$ is the kinetic energy of the ith particle. The term $V(x_j - x_i)$ is the potential energy due to the interaction of the ith and jth particles. We suppose that the potential energy depends only on the distance between the two particles, and tends rapidly to zero when the distance becomes large. Our simple global condition is then

$$\text{energy} \leq E.$$

The claim is that if a configuration of positions x_i and velocities v_i satisfies this condition, it will usually look very special, and be distinguishable from configurations corresponding to other choices of the potential V or of E. Rather unbelievable, isn't it? Well, it took some time to understand, and the people to whom credit is due are Gibbs and his followers. The details of the analysis are relatively difficult and technical, and cannot be discussed here. But there is a central idea which is simple and pretty, and which I want to explain.

I see, however, that you have an objection, and that I must face it immediately. If we have a configuration satisfying

$$\text{energy} \leq E,$$

it will also satisfy

$$\text{energy} \leq E'$$

when E' is larger than E. Therefore the configurations associated with E cannot be distinguished from those associated with E', contrary to what I just claimed, and the claim is therefore nonsense.

What saves the claim is the adverb "usually." There are many, many more configurations with energy $\leq E'$ than with energy $\leq E$. Therefore, a configuration with energy $\leq E'$ will usually not have energy $\leq E$, and cannot be confused with those low energy configurations. In more technical language, the entropy associated with E' is larger than that associated with E, and the corresponding volume in phase space (or the number of states) is very much larger.

In a sense I have just given out the simple and pretty central idea that I promised you. Let me do it again with a very simple and explicit example. I take the potential energy V to be zero, so that my global condition on the energy is now

$$\sum_{i=1}^{N} v_i^2 \leq \frac{2E}{m}.$$

To make things as simple as possible, I shall assume that my N particles are in a one-dimensional box, so that the v_i's are numbers rather than vectors, and I shall write $2E/m = R^2$. Then

$$\sum_{i=1}^{N} v_i^2 \leq R^2.$$

says that the vector in N dimensions, with components v_i, has length $\leq R$. (I have used the Pythagorean theorem.) In other words, the configurations of velocities that are allowed are the points inside a sphere of radius R in N dimensions. What is the fraction of configurations inside the sphere of radius $\frac{1}{2}R$? It is the ratio of the volumes of the two spheres: $\frac{1}{2}$ if $N = 1$, $\frac{1}{4}$ if $N = 2$, $\frac{1}{8}$ if $N = 3, \ldots, \frac{1}{1024}$ if $N = 10, \ldots$, less than one part in a million if $N = 20$, and so on. If we have many particles (i.e., if N is large), practically all configurations will be outside of the sphere of radius $\frac{1}{2}R$. Similarly, they will be outside of the sphere of radius $\frac{9}{10}R$, or $\frac{99}{100}R$.

The outcome of the argument is this: Take a sphere of radius R in N dimensions, N large; then most points inside the sphere are in fact very close to the surface. (There are exceptions, of course: the center of the sphere is not close to the surface.) We thus have an example in which a simple global condition (that a point is inside a sphere) implies—usually—a much more stringent condition (that the point is very close to the surface of the sphere). This is a rather general situation, which depends on the fact that we are willing to say *usually* rather than always. Also, we have assumed that N is large: we look at geometry in many dimensions (or at a complicated system, containing many particles).

A large part of the work of scientists is to follow a general idea (like the metaphysical idea about complicated systems expressed above) and see how far it can be justified, and when it starts to break down or become useless. In practice, this means a lot of hard work. I cannot even start giving an idea of what this hard work is,[2] but I want you to remember that it is there, and that the present nontechnical discussion is based on it. To pursue the discussion at a purely metaphysical and literary level is like driving a car blindfolded: it can only lead to disaster. Having satisfied my conscience with this warning, I can now say a little more about equilibrium statistical mechanics. It will be a bit technical and you may choose either to read the end of this chapter slowly and carefully, or on the contrary to proceed as fast as possible to the next chapter.

As we have seen, the entropy S increases (by ΔS, say) when the energy E increases (by ΔE, say). The rate $\Delta E/\Delta S$ (i.e., the derivative of the energy with respect to the entropy) is an important quantity. Let us call it *tee* or T.

Suppose now that we have a system composed of two parts, I and II (two lumps of matter in equilibrium with each other). We impose the condition

$$\text{energy} \le E.$$

As we have seen, this implies that the energy is usually almost equal to E. But there are other consequences as well: the energy of subsystem I is also nearly fixed to some value E_I) and the energy of subsystem II to some value E_{II}. How does the system choose the energies E_I and E_{II}? It just tries to maximize the sum of the entropies of system I (at energy E_I) and system II (at energy E_{II}) subject to the condition $E_I + E_{II} \approx E$. If you think about it for a moment, you will see that this makes sense: the system simply occupies as large a volume in phase space as it can, subject to the condition that its energy is fixed. But the condition that the sum of the entropies be maximum can also be expressed by saying that the *tee* of system I is equal to the *tee* of system II:[3]

$$T_I = T_{II}.$$

And this is how the concept of temperature naturally arises: *tee* may be identified with the absolute temperature, employing a conventional constant factor:

$$\text{absolute temperature} = \frac{1}{k}\frac{\Delta E}{\Delta S};$$

the factor k is Boltzmann's constant, which we have already met. Two subsystems are in equilibrium if they have the same temperature.

Notice that the temperature concept was not introduced until now, even if we have spoken loosely of cold and hot water to indicate smaller or larger total energy. Instead of beginning with experimental evidence, we started from general considerations of geometry in a large number of dimensions, and we ended up naturally with a quantity that must be the temperature. The early statistical mechanicians tried to see what a world made up of many atoms and molecules would look like—starting from scratch. Can you imagine their wonder, their excitement, their sense of power, when they found that the reconstructed world was like the one around us?

CHAPTER 20

Boiling Water and the Gates of Hell

If you don't know Russian, all books in that language will look very much the same to you. Similarly, unless you have the appropriate training, you will notice little difference between the various fields of theoretical physics: in all cases what you see are abstruse texts with pompous Greek words, interspersed with formulas and technical symbols. Yet different areas of physics have very different flavors. Take for instance special relativity. It is a beautiful subject, but it no longer has mystery for us; we feel that we know about it all we ever wanted to know. Statistical mechanics, by contrast, retains its awesome secrets: everything points to the fact that we understand only a small part of what there is to understand. What are these awesome secrets? The present chapter will describe a couple of them.

One puzzling natural phenomenon is the boiling of water, and the freezing of water is no less mysterious. If we take a liter of water and lower the temperature, it is not unreasonable that it should become more and more viscous. We may guess that at low enough temperature it will be so viscous, so stiff, as to appear quite solid. This guess about the solidification of water is wrong.[1] As we cool water we see that at a certain temperature it changes to ice in a completely abrupt manner. Similarly, if we heat water it will boil at a certain temperature, i.e., it will undergo a discontinuous change from liquid to water vapor. The freezing and boiling of water are familiar examples of *phase transitions*. These phenomena are in fact so familiar that we may miss the fact that they are very strange indeed, and require an explanation. Perhaps one could say that a physicist is a person who does *not* consider it obvious that water should freeze or boil when its temperature is lowered or raised. What does statistical mechanics tell us about phase transitions?

According to our general philosophy, imposing a global condition (in this case fixing the temperature) has the result that all kinds of things about the system are uniquely specified (*usually*). Given a snapshot of the configuration of atoms in helium at 20° C, you should be able to distinguish it from a snapshot corresponding to another temperature or another substance, in the same way as you distinguish a van Gogh from a Gauguin at a glance. The "cluster of probabilistic features" changes with temperature, and the change is usually gradual. In the same way, the style of a painter might gradually change as the artist gets older. And then the unexpected occurs. At a certain temperature, instead of gradual change you have a sudden jump—from helium gas to liquid helium, or from water to water vapor or to ice.

Can one easily recognize ice from liquid water in a snapshot of the molecules? Yes. Ice is crystallized (think of a snowflake), and the directions of the axes of the crystal can be seen in the snapshot as statistical alignments of molecules in certain directions. In liquid water, by contrast, there are no preferred directions.

So, here is a problem for theoretical physicists: prove that as you raise or lower the temperature of water you have phase transitions to water vapor or to ice. Now, that's a tall order! We are far from having such a proof. In fact, there is not a single type of atom or molecule for which we can mathematically prove that it should crystallize at low temperature. These problems are just too hard for us.

If you are a physicist, you won't find it unusual to be confronted with a problem much too difficult for you to solve . . . There are ways out, of course, but they require that your relation to reality be altered in one way or the other. Either you consider a mathematical problem analogous to the one you cannot handle, but easier, and forget about close contact with physical reality. Or you stick with physical reality but idealize it differently (often at the cost of forgetting about mathematical rigor or logical consistency). Both approaches have been used to try to understand phase transitions, and both approaches have been very fruitful. On one hand it is possible to study systems "on a lattice" where the atoms instead of moving freely can be present only at some discrete sites. For such systems one has good mathematical proofs that certain phase transitions occur.[2] Or one can inject new ideas into the idealization of reality, like Wilson's ideas of *scaling*, and obtain a rich harvest of new results.[3] Still, the situation is not quite satisfactory. We should like a general conceptual understanding of why there are phase transitions, and this, for the moment, escapes us.

To show the power of the ideas of statistical mechanics, I shall now jump from boiling or freezing water to something totally different: black holes.

If you shoot a bullet up in the air, it will fall back after a while, because its speed is insufficient to overcome gravity, i.e., the attraction of the bullet by the earth. But a very fast bullet, with speed greater than the so-called *escape velocity*, would leave the earth forever if we ignore small details like the slowdown by air friction. The escape velocity for some celestial bodies is less than that for the earth, and for others it is greater. Suppose you are on a small, massive celestial body, for which the escape velocity is greater than the speed of light.

Then anything you try to send up, including light, will fall down. You cannot send any message to the outside world; you are trapped. The sort of celestial object on which you are is called a *black hole*, and should be signaled with the same warning that, according to Dante, is written over the gates of Hell: *Lasciate ogni speranza, voi ch'entrate*. Abandon all hope, you who enter . . .

In fact, my description of a black hole was a bit naive: red lights start flashing and sirens start howling in the mind of a physicist when you mention a "speed greater than the speed of light." When we discuss gravity and the speed of light at the same time, the physical theory that we should use is *general relativity*. According to Einstein's theory of general relativity, black holes actually exist and can rotate. They are formed when a large amount of matter is put in a small region of space; they attract and swallow anything that happens to be nearby. Astrophysicists do not have ironclad proof that they have seen black holes, but they think they have. In particular, very powerful sources of radiation present at the center of galaxies, and also quasistellar objects ("quasars"), are believed to be associated with very massive black holes. The radiation is not emitted by the black hole itself, which can in principle not emit anything, but by the surrounding regions. Those regions, if we believe the astrophysicists, are extremely unpleasant places, as unhealthy as the gates of Hell. Indeed, if a physicist is in charge of Hell, the latter will probably look like a massive black hole. Suppose 1E9 (one billion) solar masses have collapsed to form a rotating black hole. There will be an *accretion disk* of matter spiraling down toward the hole. This matter, hot and ionized, forms a conducting plasma and will normally carry some magnetic field with it. One can try to figure out the dynamics of the infalling matter, magnetic and electric fields, electric currents, and so on. The results are forbidding. Estimated voltage drops on the order of 1E20 volts (twenty zeros!) develop around the hole. Electrons are accelerated by these potential differences and collide with photons (particles of light), which hit other photons, and produce an inferno of electron-positron pairs. At least, this is one view of what is going on. There is no general agreement as to the details, but the overall picture is that of a region of about the size of our solar system radiating a huge amount of energy. You remember that energy and mass are equivalent, via Einstein's famous $E = mc^2$. The energy output in this case would be on the order of 10 solar masses per year, a hideously large amount any way you look at it.

But theoretical physicists are not easily impressed, and they continue to ask questions like the following. Suppose that a black hole, instead of being in the middle of an accretion disk of infalling matter, sits quite alone in a complete vacuum. What would we see of such a pure black hole? Would it produce any radiation? According to the classical ideas of general relativity, a pure black hole would have gravitational effects: it would attract distant matter, and a rotating black hole would also make the matter turn. A black hole might also have electric charge, which we shall ignore for simplicity. Apart from that, pure black holes are very much alike. Two black holes with the same mass and the same rotation (i.e., the same angular momentum) are indistinguishable. Whether the black hole was produced out of hydrogen or out of gold makes no difference. The hole has forgotten its origins (apart from mass and angular momentum), and a physicist will refuse to speak of a hole made of hydrogen or of gold. Furthermore, according to general relativity, the hole does not produce any radiation.

Among the people who looked into the problem of black holes was the British astrophysicist Stephen Hawking, and he was not satisfied with the answer concerning the absence of radiation. The verdict of general relativity is clear, but does not take quantum mechanics into account. (In fact, we don't have a completely consistent theory unifying quanta and general relativity.) Why would quantum mechanics be important for this problem? The reason is that according to quantum theory, the ''vacuum'' cannot be completely empty. If you look at a very small region of vacuum, the position is rather precisely known, and therefore the Heisenberg uncertainty relations assert that the velocity (more precisely: the momentum) must be rather uncertain. This means that there must be *vacuum fluctuations* in the form of particles flying by at high speed.[4] I know that this argument sounds like a swindle, but it is the best way to put into words what the mathematical formalism would express in more coherent manner. Normally, vacuum fluctuations become insignificant if you look at a larger region of vacuum. But what if the vacuum is submitted to the intense gravitation near a black hole? Well, according to Hawking's calculations, some of the particles constituting the vacuum fluctuations fall into the black hole, and others escape in the form of radiation. In fact, the black hole emits electromagnetic radiation (light) just like any lump of hot matter, and one can thus speak of the temperature of a black hole.

Hawking's result was received at first with considerable skepticism by physicists, but gained acceptance as the calculations were redone and new insight into the problem was gained from various sources.[5] Perhaps it should be said right away that massive black holes have very low temperatures, and that their Hawking radiation is quite undetectable. This radiation has, however, a tremendous theoretical interest, of which I shall now try to give you a glimpse.

Let us go back to the fact that entropy cannot decrease (the so-called second law of thermodynamics). It would seem that you can contradict this fact by dumping stuff with a lot of entropy down a black hole. (There will be a little increase in mass, but otherwise, the hole will forget what you dumped into it.) It is, however, possible to rescue the second law of thermodynamics by giving the black hole an entropy (depending

CHAPTER 21

Information

Dipped in your own blood, the pen screeches on the parchment. You have just signed a pact with the Devil. You promise him your soul after death if he will give you wealth, and all that goes with it, during your lifetime. How will he keep his part of the agreement? Perhaps he will let you know the coordinates of a hidden treasure, but that is a bit old-fashioned. More conveniently, he will tell you in advance the results of horse races, and make you moderately affluent. If you are really greedy, he may give you forecasts of the stock market. Knowledge is what the Devil has to offer. In all cases, what you get in payment for your soul is knowledge, information: coordinates of a treasure, names of winning horses, or lists of stock values. Information is what will make you rich, loved, and respected.

Here is another example of the power of information. Suppose that some alien species wants to eradicate mankind from the earth without damaging the environment. One way they can proceed is by use of a suitable virus. What they want is a virus as lethal as the AIDS virus, but also easily transmissible and quick-acting like some new strain of the common cold virus. They want something that leaves us no time to devise strategies, arrange vaccines, and the like.

At this time, the virus needed to eradicate mankind presumably does not exist on the surface of the earth. But it could be produced with suitable technology. What the alien species wants is a blueprint—information again. In the case of AIDS, the needed information is essentially contained in the particular sequence of bases that encodes the genetic information of the virus. This sequence is a message written with a four-letter alphabet (A, T, G, C),[1] and it contains 9749 letters or thereabouts. It is a rather short message. Probably there is a similar message coding for a virus that is lethal, quick, transmissible, and capable of wiping us all out. This message would spell the doom of mankind, and could be printed on a few pages of the book in your hands.

I personally wouldn't worry too much about unfriendly aliens. Crazy heads of state and fanatic governments seem a greater menace. They would have no trouble finding scientists with confused idealistic ideas or conscientious unimaginative technicians to implement the most insane scheme. Perhaps this is how the history of mankind will terminate.

I have only one comforting thought to propose to you in this respect. If the unfriendly aliens or crazy scientists have to rely on sheer luck to find the blueprint of the ultimate virus, then we are very safe indeed. The number of messages with about ten thousand letters written in a four-letter alphabet is just too large to sift through: there are many more such messages than there are grains of sand on all the beaches of the Galaxy, many more in fact than there are atoms in the whole known universe. In brief, no one can expect to guess correctly a message that is ten thousand letters long.

The length of a message gives an indication of its *information content*, and tells us how hard the message is to guess. Let us try to get a more precise definition of the information content of a message. The length is important, but the alphabet also certainly plays a role: you can replace the four letters A, T, G, C by the two symbols 0, 1 at the cost of translating one letter by a pair of symbols: A = 00, T = 01, G = 10, C = 11. The translated message has twice as many symbols as the original one but has the same information content. Or you could code pairs of successive letters A, T, G, C by sixteen letters of the alphabet a, b, c, . . . , p, making the message only half as long, but still containing the same information.

If you have a message in English you can compress it by omitting the vowels, and the message usually remains understandable. This means that written English is redundant: more is spelled out than is needed for understanding. However, to decide what the information content of a message is, you have to know if it is written in English or French or some other language. More generally, you have to know what the allowed messages of a certain length are. If you have the list of the allowed messages, you can number them, and specify any one by giving its number. This coding of possible messages by their number has no redundancy, and therefore the length of the coding numbers is a good measure of the information content of the messages. The following is thus a reasonable definition:

information content = number of digits of the number of allowed messages.

This definition refers to a class of allowed messages rather than to a single one (another point of view is possible and will be discussed in a later chapter). Some adjustment of the definition is needed when the different messages do not all have the same probability, but we need not bother about this here.[2]

By reference to what we said about entropy, we might also write

$$\text{information content} = K \log(\text{number of allowed messages}).$$

Mostly, the information content is expressed in *binary digits* or *bits*. This means that you translate the message into an alphabet with two "letters" 0 and 1, and then measure its length (or take $K = 1/\log 2$ in the above formula).

In a paper published in 1948,[3] the American scientist Claude Shannon single-handedly created *information theory*. The theory in question deals with a very important practical problem: transmitting information efficiently. Suppose you have a source producing a constant stream of information (a politician delivering a speech, or your mother-in-law chatting over the telephone—it need not be meaningful information). You may consider the stream of information as a succession of messages of some given length, written in English, and produced at a certain rate. Your job, as a technician, is to transmit these messages over a certain line. The line may be an old-fashioned telegraphic cable, or a laser beam aimed at some distant space station. The line has a certain *capacity*—the maximum number of binary digits (or bits) that it can transmit per second. If your source of information produces more bits per second than the capacity of your line, you cannot transmit the message (at least not at the rate at which it is produced). Otherwise you can, but you may have the problem of getting rid of some redundancy of the original message by coding it properly. (This is called *data compression*; the message can be compressed if it is redundant, but the information is not compressible.)

Another problem that could arise is noise over the line. This you may overcome by increasing the redundancy of the message in an appropriate manner. Here is what you do. When you code the message, you introduce extra bits of information that allow you to check when the noise has changed a letter, and again other bits allowing you to make corrections. In other words, you use an error-correcting code. If the capacity of your line is high enough, and the noise low enough, you may beat the noise with error-correcting codes. More precisely, you can arrange that the probability of incorrect transmission is arbitrarily low. This requires proofs, of course, and the theory of error-correcting codes is difficult, but the basic ideas are simple.

The definition of information was modeled after that of entropy, the latter measuring the amount of randomness present in a system. Why should information be measured by randomness? Simply because by choosing one message in a class of possible messages you dispel the randomness present in that class.

Information theory has been a remarkably successful scientific discipline, both in its mathematical developments and in its practical applications. As in the case of physical theories, it must however be understood that information theory deals with idealizations of reality and leaves out certain important features. The source of information is supposed to produce a random sequence of allowed messages (or an infinitely long message with certain statistical properties). It is not required that the messages be useful or logically coherent, or that they have any meaning at all. Saying that a message has high information content is the same thing as saying that it is extracted from a large class of allowed messages, or that it is very random. Some of this randomness may correspond to useful information, and some may be trash.

Let us discuss an example: musical melodies. We leave aside various details, and consider melodies as messages in which the alphabet is a musical scale. We may try to find the information content (or the randomness) of a melody by studying the frequency of the various notes, and the statistics of intervals between successive notes (this is a standard procedure in information theory).[4] As mentioned before, older music mostly uses small intervals, and therefore few of them. In more recent times, an increasing variety of intervals frequently occurs. From this one can conclude that (in Western classical music) there has been a gradual increase in the information content, or randomness, of musical melodies.[5] This is an interesting conclusion, but one that has to be taken with a grain of salt. Indeed, there is more to a musical melody than the statistics of successive intervals. A musical piece has a beginning and an end, and quite a bit of structure in-between. This structure does not just correspond to correlations between successive notes (the statistics of intervals) but also to long-range correlations (correlations over the whole length of the piece), which are not captured by the usual information-theoretic descriptions.

Also, information in a melody may be inventive and original, or meaningless and dull. If you put musical staves over a map of the sky and mark notes at the positions of stars, you obtain ''celestial music,'' which has a lot of information, but this does not mean that it is very good music.

The information content of a work of art is an important notion (it could be defined for paintings as well as for poems or melodies). This does not mean that high quality is equivalent to a lot of information or to very little information. Probably one cannot speak of art unless there is a minimum of information, but some artists have tried very low values. By contrast, the information content of many works of art (paintings or novels) is huge.[6]

Perhaps you are, at this point, getting mildly irritated that I am discussing the information content of messages and sweeping the problem of their meaning quietly under the rug. More generally, you may feel that scientists systematically address the more formal and superficial questions and leave out the essential ones. The answer to this criticism is that science emphasizes good answers (and if possible, simple answers) rather than deep questions. The problem of meaning is obviously deep and complex. It is tied among other things to the question of how our brain works, and we don't know too much about that. We should thus not wonder that today's science can tackle only some rather superficial aspects of the problem of meaning. One of these superficial aspects is information content in the sense discussed in the present chapter, and it is remarkable how far this carries us. We can measure quantities of information the same way we measure quantities of entropy or electric current. Not only does this have practical applications, but it also gives us some insights into the nature of works of art. Of course we would like to ask more ambitious questions, but in many cases it is apparent that these more difficult questions are too hard for us to answer. Think of musical melodies; they are messages that we feel we understand, yet we are quite incapable of saying what they mean. The existence of music is a permanent intellectual scandal, but it is just one scandal among many others. Scientists know how hard it is to understand simple phenomena like the boiling or freezing of water, and they are not too astonished to find that many questions related to the human mind (or the functioning of the brain) are for the time being beyond our understanding.

CHAPTER 22

Complexity, Algorithmic

Science progresses through the creation of new concepts: new idealizations in physics, new definitions in mathematics. Some of the newly introduced concepts are found, after a while, to be unnatural or unproductive. Others turn out to be more useful and fundamental than anticipated. *Information* has been one of the more successful concepts of modern science. Among other things, information allows us to approach the problem of *complexity*.

We are surrounded by complex objects, but what is complexity? Living organisms are complex, mathematics is complex, the design of a space rocket is complex. What do these things have in common? Well, probably that they contain a lot of information that is not easy to come by. We are as yet unable to produce living organisms from scratch, we have a hard time proving some mathematical theorems, and the design of a space rocket requires a lot of effort.

An entity is complex if it embodies information that is hard to get. We have not said what "hard to get" means, and therefore our definition of complexity has no sharp meaning. In fact, the natural languages that we use in everyday life (here, English) allow us to give wonderfully vague definitions like the one above. This is more a blessing than a nuisance, really. But if we want to do science we have to be more precise, more formal. As a consequence, there will be not one but several definitions of complexity, depending on the background in which we place ourselves. For instance, a serious discussion of the complexity of life must include as background the physical universe in which life develops. But there are also concepts of complexity that can be developed with a purely mathematical background. I shall now discuss one such concept, that of *algorithmic complexity.*

In brief, an algorithm is a systematic way of performing a certain task. We have all learned the algorithm for multiplying two integers, for instance. An algorithm always works on an input message, like "3 × 4" (written with the symbols 0, 1, 2, . . . , 9, ×), and returns an output message, like "12." Of course, multiplications are best handled nowadays with the use of a computer, and one may define an algorithm as the task performed by a computer (with a suitable program in it). What we mean by computer is actually a slightly idealized machine that has an infinite memory at its disposal. (We don't want to restrict the definition of algorithms just because commercial computers cannot enter in memory a number with 1E100 digits.)

The British mathematician Alan Turing invented and described precisely a computer that is well suited to the theoretical study of algorithms, although it would be remarkably inadequate for implementing them in practice. The *Turing machine* has a finite number of *internal states*: some so-called active states and one halting state.

The machine does its work on an infinite ribbon of paper divided into a succession of squares. (This ribbon serves as memory.) On each square of the ribbon is marked a symbol from a finite alphabet, one of the symbols being *blank*. The Turing machine operates at successive moments of time in a completely predictable way. If it is in the halting state, it does nothing at all. Otherwise, the dimwitted machine reads the square over which it is sitting and then, depending on its internal state and on what it has just read, does the following things:

(a) it erases what was written and writes something else (or the same thing) in the square,
(b) it moves one square to the left or one square to the right,
(c) it changes to a new internal state.

The machine then starts another cycle depending on what it reads in the new square and what its new internal state is.

The initial state of the tape contains a finite message, which is the input message (the rest of the tape is blank, i.e., consists of squares marked with the "blank" symbol). The machine is started at one end of the message, and things are arranged so that when it halts it has written a message of its own, which is its answer. The answer may be Yes or No, or it may be a digit, or it may be a longer message. One can set up a Turing machine to add or multiply integers. Indeed, a Turing machine can do much more than perform multiplications: any task that can be done by a computer, can also be performed by a suitable Turing machine. And in fact one does not need an infinite number of machines for different tasks, because *there exists a universal Turing machine*. To implement a particular algorithm on that machine one has to write on its ribbon an input message that contains both the description of the algorithm and the particular data that one wants to handle.[1]

To summarize: An algorithm is something that can be implemented on a computer, and we might as well use a very primitive kind of computer called a Turing machine. Given a certain task, there may be efficient and inefficient algorithms to perform it, depending on the number of cycles of the Turing machine needed to get an answer. The *algorithmic complexity* of a problem depends therefore on the availability of efficient algorithms to handle the problem. The accepted definition of an efficient algorithm compares the length L of the input message (i.e., its information content) and the time T (number of universal Turing machine cycles) needed to get an answer. If

$$T \leq C(L+1)^n,$$

where C and n are some constants, we have a *polynomial time algorithm*. (The reason for this name is that $C(L+1)^n$ is a polynomial in L.)

A polynomial time algorithm is considered efficient, and the corresponding problem is called *tractable*. If $n = 1$, the time taken to implement the algorithm is at most proportional to the length of the input (plus one); if $n = 2$, it is at most proportional to the square of the length of the input (plus one); and so on. One can prove that the definition of tractability does not depend on the particular universal Turing machine that is used. As an example, let us consider the problem in which the input message is an integer and we wish to know if this integer is divisible by 2, or by 3, or by 7. It will be no surprise to you that these are tractable problems (and you may have learned in school the efficient algorithms to handle them).

Basically, modern computers are universal Turing machines (they are only a bit deficient in not having an infinite memory). Computer scientists therefore like to know what the tractable problems are. But the discovery of an efficient algorithm can be rather difficult. This has been the case, for instance, for *linear programming*, which was shown only in recent years to have a polynomial time algorithm.[2] In linear programming, technically, the question is to find the maximum of a linear function on a convex polyhedron. The minimax theorem in the theory of games leads to such a question, and there are many problems of resource allocation that also lead to linear programming questions. In this case, therefore, the proof of tractability may have important practical consequences.

Efficient algorithms are not always available, however. Suppose the only way we know to handle a problem involves a case-by-case search among all messages of length L in a binary alphabet. This will take a time

$$T \geq 2^L.$$

Here, the estimated minimum time required for solving the problem is multiplied by 2 whenever the length L is increased by 1. We have seen examples of such *exponential growth* in earlier chapters, and convinced ourselves that it quickly gives very large numbers. An exponential time algorithm is thus not very practical. In general, a problem for which a polynomial time algorithm does not exist is considered *intractable*.

So, what are examples of intractable problems? And why are they intractable? I suggest that you pose those questions to a theoretical computer scientist, if you count one among your friends. Allow a few hours for the answer, and try to have a blackboard at your disposal. It is not that it is so difficult to explain, but it is, let us say . . . a bit technical. It is also absolutely fascinating. Your friend will define *NP complete* problems,[3] *NP hard* problems, and explain to you that such problems are believed to be intractable. It would be fantastic if one could prove that NP complete (or hard) problems are intractable. It would be even more fantastic if one could prove that they are tractable . . .

Perplexed? Well, all that I can reasonably try to do here is give succinct indications on these topics, and examples of problems that are believed to be intractable.

A popular example is the problem of the traveling salesman. You are given the distances between a certain number of cities, and you are allowed a certain total mileage. (The distances and total mileage are integers, counted in miles or some other unit.) The question is whether there is a circuit joining all cities with length at most the total allowed mileage. This is a Yes or No question. If a certain circuit is proposed, it is rather easy to check whether it satisfies the total allowed mileage condition.

But to test all possible circuits one by one when there are many cities would be intractable. This is an example of an NP complete problem.

In general, NP complete problems require Yes or No answers, and have the feature that one can verify the existence of a Yes answer in polynomial time. (There is an asymmetry between the Yes or No answers, because one does not say that a No answer can be verified in polynomial time.) Let "Problem X" be your favorite Yes or No problem. Suppose that Problem X becomes tractable if you have free access to solutions of the traveling salesman problem, and that the traveling salesman problem becomes tractable if you have free access to the solutions of Problem X; then Problem X is said to be NP complete. In spite of extensive search, no polynomial time algorithm has been found to solve NP complete problems, and it is generally believed that none exists. But this has not been proved.

It is convenient to introduce NP hard problems, which are just as hard as NP complete problems but do not require Yes or No answers. Here is an example: the *spin glass problem*. The input message is an array of numbers $a(i,j)$ equal to $+1$ or -1, where i and j go from 1 to some value n (for instance, from 1 to 100, in which case there are 10,000 numbers ± 1 in the array). You are asked the maximum value of the expression

$$E = \sum_{i=1}^{n} \sum_{j=1}^{n} a(i,j)\, x(i)\, x(j),$$

where $x(1), \ldots, x(n)$ are allowed the values $+1$ or -1. You thus have to add n-squared terms, each equal to $+1$ or -1, and make the result maximum. Perhaps you can't believe that this is an intractable problem, and perhaps it isn't, but nobody has found an efficient algorithm for solving it. (Note that the input message has n-squared bits, and that a case-by-case search requires considering 2^n cases.) The spin glass problem is the prototype of a family of problems that arise in the physics of *disordered systems*.[4] (What is disordered is the "interaction" $a(i,j)$ between the sites i and j.) The problem of making E maximum is like the problem of finding the highest peak of a chain of mountains (see Figure 22.1). In the case of the figure, this is easy, because x varies on a line (i.e., x is one dimensional). In the spin glass problem, the geometry of peaks and valleys is n dimensional . . . and intractable (even though, for each of the n dimensions, only the two values $+1$ and -1 are possible).

Let us make an idealization—or, better said, a metaphor—of the problem of life. According to the metaphor, the problem of life is to find a genetic message $x(1) \ldots x(n)$ that gives a very large value to a complicated expression like E above. According to what we have just said, this may be a very difficult problem. There are indications that the above metaphor of life is perhaps not so far wrong.[5]

The idea of algorithmic complexity may also serve as a metaphor for the difficulty of proving mathematical theorems, or designing a space rocket. We shall see, however, that proving theorems leads us to deeper layers of complexity than the NP complete problems: deeper, more obscure, and more revolting.

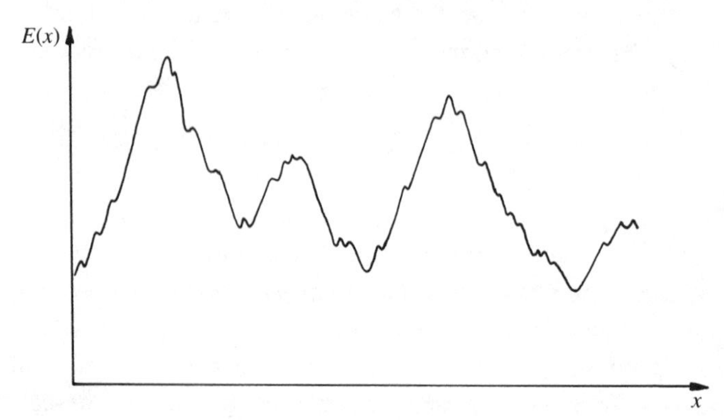

Figure 22.1. What is the largest value of $E(x)$?

CHAPTER 23

Complexity and Gödel's Theorem

In 1931 the Austrian-born logician Kurt Gödel published what is probably the single most profound conceptual result obtained by mankind in the twentieth century. I remember seeing Gödel at the Institute for Advanced Study in Princeton during the sixties and early seventies. He was a small man, yellowish and emaciated, and he wore cotton plugs in his ears. Here is a typical story I heard about him.[1] A visiting colleague was allowed the use of Gödel's office while the latter was away. Said colleague, upon leaving, put a note of thanks on the desk, saying that he regretted not having seen Gödel, and expressing the hope that he would have a chance to get to know him more intimately on a later occasion. Some time later he got an envelope in the mail from Gödel. The envelope contained his own note with the sentence *I hope to have a chance to get to know you more intimately on a later occasion* underlined by Gödel, who had added in pencil the question: *Exactly what do you mean?*

Kurt Gödel died in 1978 from self-inflicted starvation. He apparently thought that people were trying to poison him, or something, and he refused to eat.

If you count the suicides of Ludwig Boltzmann and Alan Turing (who was a homosexual at a time and place where this was not accepted), you may come to the conclusion that scientists are a rather suicidal bunch. This would be a completely wrong conclusion. Most scientists are in fact quite normal, normal often to the point of being dull and unimaginative. And I don't think I shall be contradicted when I say that also in their scientific work many are dull and unimaginative. Even their obituaries are often dull and stereotyped, lamenting their untimely death, pointing to their active role in the synagogue or church community, and also celebrating their "infectious enthusiasm" and similar nonsense. (Infectious enthusiasm is a distressful condition, often diagnosed only *post mortem*.)

But let us return to Kurt Gödel. Whatever his problems were, at least he did not suffer (and make others suffer) from infectious enthusiasm.

To understand Gödel's discovery, it is perhaps a good idea to reflect on a psychological setup characterized by order, parsimony, and stubbornness, common among scientists (particularly mathematicians) and useful to them. This setup has been related by Sigmund Freud to a predisposition to obsessional neurosis and the so-called anal-sadistic stage of evolution of the libido.[2] In any case, such a psychological disposition makes it natural to try to present mathematics, and mathematical deduction, in a form that is as clean and orderly as possible. The great dream, then, is to base mathematics on sharply defined rules of inference, and a finite number of completely explicit fundamental assertions called axioms. This dream grew from Euclid the Greek (around 300 B.C.) to David Hilbert, the great German mathematician (1862–1943), and led to a progressive formalization of the whole of mathematics. The arithmetic of integers was formalized rather early, and the culmination of the great dream of mathematicians was the following hope: that for every meaningful assertion about integers, one could decide in a systematic manner if it was true or false. This is the hope that Gödel shattered.

Gödel showed that if you fix the rules of inference, and any finite number of axioms, there are meaningful statements that can be neither proved nor disproved. More precisely, suppose that the axioms accepted for the integers are *noncontradictory*, i.e., suppose that by application of the rules of inference you can never prove that an assertion is at the same time true and false. Then there are true properties of the integers[3] that cannot be derived from the axioms. And if you accept any such property as a new axiom, there will remain other unprovable properties.

Gödel's incompleteness theorem has played a pivotal role in our understanding of the foundations of mathematics. At first it was a great shock. Then it led to a progressive change in the systems of beliefs of mathematicians. Simultaneously, the difficult proof of the theorem was simplified. This simplification came from the introduction of new concepts, in part by Gödel, in part by others (the Turing machine is a relevant example). Altogether, the discovery of the incompleteness theorem has led to a progressive change of the landscape of mathematics. And the result is that the incompleteness theorem now appears rather natural, and in fact somewhat trivial. The earlier great hope was that some finite set of true assertions (called axioms) would form a basis from which all true assertions about integers could be derived. We now know that *the set of all properties of integers* (i.e., the set of all true assertions about them) *does not have a finite basis*. We also have an intuitive understanding of why a finite basis cannot exist, and this is again based on *information*, as I shall now indicate.

We saw earlier how one could define the information content of a message, given the family of messages to which this belongs. In particular, if all the messages consisting of symbols 0 and 1 are accepted, a sequence of a million 0's has an information content of one million bits. Another idea, proposed by Solomonoff, Kolmogorov, and Chaitin,[4] is to consider the length (in bits) of the shortest computer program that will produce the message of interest as an output. In the present case, the program would be something like "print one million 0's," and its length would be much shorter than one million. The quantity thus defined has been called *algorithmic information*, or *Kolmogorov-Chaitin complexity*. It is a complexity in the sense that it measures how hard it is to produce the message (how hard in the sense of program length, in bits, not in the sense of computing time). Depending on the choice of computer, slightly different definitions are possible, but one can for instance use a universal Turing machine.

If the message 'blah blah blah . . .' has a million bits, its K.-C. (Kolmogorov-Chaitin) complexity cannot be much more than one million, because you can print it using the program: "print 'blah blah blah . . .' " Also, if a message has a million bits, its K.-C. complexity is usually not much less than one million. (It makes sense: most messages cannot be compressed to 10 percent of their original length, for instance; only a very small fraction can.) These are fairly easy remarks.

Let me now turn to a more difficult problem: given a certain message, determine its K.-C. complexity. You were yawning, weren't you? You don't care about K.-C. complexity? You are bored? Well, I shall take advantage of this lapse in your vigilance, give you bad advice . . . and in a few minutes you will be swamped in logical paradox, and asking for mercy.

How do we determine the K.-C. complexity of the message 'blah blah blah . . .' one million bits long? Well, we make a list of all programs not much longer than one million bits, insert them one by one into our computer, and watch the output. The length of the shortest program with output "blah blah blah . . ." is the K.-C. complexity of that message. Nothing easier. It may take too long to be done easily in practice, but you don't see any reason why it couldn't be done in principle. Do you?

Well, well, well! While we are at it, we may ask our friendly computer to print the message that comes first in alphabetic order among those that have a K.-C. complexity of at least one million. I leave you to find out how to define alphabetic order in the present context. I leave you also the task of writing the "superprogram" that prints the first message (in alphabetic order) with a complexity of at least a million. This superprogram should be fairly short (it checks a finite number of programs and prints one output). If you are any good at programming, your superprogram should have fewer than a million bits . . . and there you are, up to your neck in paradox, and asking for mercy: with a program that contains fewer than one million bits, you have defined a message with a K.-C. complexity of at least one million, in contradiction to the definition of K.-C. complexity.

What did you do wrong? Logicians will tell you that your mistake was to sit by the computer after entering a program, and imagine that in due time it would produce an output. A Turing machine may after some time halt and produce an output, or it may never halt, *and you don't know it in advance*. You shouldn't expect too much from a Turing machine. In particular, you should not expect to know whether it will ever halt on a given input; there is no algorithm to decide that. In fact, there is also no algorithm to decide the K.-C. complexity of messages—this is one aspect of Gödel's theorem, as discovered by Chaitin.

What Chaitin showed is that assertions of the type "The message 'blah blah blah . . .' has a K.-C. complexity of at least *N*" are either false, or unprovable when *N* is sufficiently large. How large is sufficiently large? That depends on the axioms of your theory. Your axioms contain a certain amount of information (depending on their total length), and you cannot prove that 'blah blah blah . . .' contains more information than the axioms you are using. It makes sense, doesn't it? And in fact it is not very hard to prove.[5]

There is a lot more that could be said about Gödel's theorem, but as I do not wish to drown myself (and you) in technicalities, I shall only add a couple of remarks.

You may be upset because I said that Gödel's theorem was about properties of integers, and then I discussed instead the complexity of messages. Actually, one can translate logical statements (for instance, concerning the complexity of messages) into properties of integers. This is a game that was started by Gödel, and has culminated in what is known as the solution of "Hilbert's tenth problem."[6] It does not matter, therefore, that we did not explicitly speak of properties of integers.

The heart of Gödel's theorem, the way we looked at it, is that we don't know if a Turing machine will halt or not when we input a certain program. For programs of a certain length, the machine will either function up to some maximum time and then halt, or function forever and never halt. If we knew the maximum halting time of our Turing machine for every program length, we could decide for which programs the machine will halt and for which programs it will not halt. (Just let the machine run up to the maximum halting time for the given program length; if it has not halted by then it never will.) But the heart of the matter is that we do not know the maximum halting time. And we cannot know it because it grows faster than any computable function of the program length—faster than a polynomial, faster than an exponential, faster than an exponential of an exponential . . .

We decided in the preceding chapter that a problem was intractable if you could not solve it in polynomial time (polynomial, that is, with respect to program length). We see how much more intractable certain mathematical problems are. We were wondering about the complexity of things, and Gödel's theorem tells us that the arithmetic of integers is already as impossibly complex as we can imagine.

A last question now: What does all this have to do with the

topic of this book? What does Gödel's theorem have to do with chance? We know that one can forever produce new properties of integers, independent of those already known, but are these in some sense random properties? The answer is indeed yes, and one can produce a sequence of properties of integers that are randomly true or false (this was done explicitly by Chaitin).[7] In other words one can, based on properties of integers, define a sequence of binary digits that are 0 or 1 independently and with probability ½. What this means is simply that no amount of computing power will give you any betting advantage (in the average) in predicting the next digit (i.e., the sequence is in fact quite uncomputable).

Strange world that we live in, isn't it? At least that is the view of the world that logicians give us at this moment. And we are getting used to it. The view may change again, and look even wilder . . . and again we shall get used to it after some time.[8]

CHAPTER 24

The True Meaning of Sex

We are getting close to the end of this book, and perhaps you regret that you haven't had a chance to take more initiative. It is true that apart from shaking your head and mumbling things, your attitude has been somewhat passive. Let us change that. I suggest that you now engage in a noble and fulfilling enterprise: creating life.

We shall assume that You have already created the Stars, the Galaxies, and all That. Writing a few Equations on a Piece of Paper is all that You had to do to create the Universe, and You shall now send Your Message to the Universe, and put Life into It.

If you don't mind, I shall now suppress the capitals and look at you and your message of life with a cold scientific eye, and in lower case. A basic fact to keep in mind is that your message of life has to prevail in the face of a lot of randomness. Indeed, classical chaos, quantum uncertainty, and even Gödel's theorem conspire to introduce chance in the universe you have created. How will this affect your message?

We discussed earlier a spin-glass model as a metaphor for life. The idea is that there is a function

$$E(\text{message})$$

that your message should try to make maximum (or at least reasonably large). We may assume that your message has to reproduce itself, and that the function E is related to the probability of reproduction of a message identical with (or similar to) the original one.[1] The function E reflects all that your message knows of the universe, and in particular it reflects the randomness of the universe.

The spin-glass problem (of making E maximum) is NP hard, as we saw in the chapter on algorithmic complexity. You won't bother to try to solve it exactly, or to solve it yourself. You will let your message take care of itself, which is a sensible thing to do, hoping that by trial and error it will reach a high value of E. Your message, in effect, is a genetic message, endowed with the ability to reproduce. Trial and error thus mean random mutation and then selection, and we are back to a relatively orthodox view of life. Mutation and selection is of course also a way to attack the spin-glass problem, but one then speaks of a *Monte Carlo method* (so called because chance plays a role in it, as in the casino). Whatever the name, we see that the trial and error method will probably lead you step by step to larger values of E, but not necessarily to the absolute maximum. Looking at Figure 22.1, you see that if you start climbing the wrong mountain step by step you will reach the top of that mountain, not the top of the highest mountain. The mutation and selection method is thus a good way to develop life, but in general it does not give optimal results.

In fact, the longer your genetic message, the more unsatisfactory the pure Monte Carlo approach is. Indeed, the information contained in your genetic message will be lost rather rapidly owing to mutations in the successive generations, unless you keep the mutation level rather low.[2] But this means that the slow process of mutation and selection will lead you only to the top of a little mountain in Figure 22.1, and that you are very unlikely ever to reach the high summits.

Creating life, as you see, leads to no end of trouble. What can you do now? A good idea would be to look at the function

$$E(\text{message})$$

that contains all the complexity of the universe, as seen from the point of view of your message of life, and try to figure out how random it is. Is there not some regularity in this function that could be made use of? Is the universe totally meaningless, or does it have some structure? Fortunately, there is some regularity in the universe, and it expresses itself even at the level of your message. What happens is that you can cut your message into pieces, or sentences, that have some meaning of their own:

$$\text{message} = (\text{sentence } A, \text{sentence } B, \text{sentence } C, \ldots).$$

The sentences A, B, C, and so on may also be called genes, and their meaning is that they code for, say, different enzymes. But I don't want to go down to the level of genetic machinery. Rather, it is important to understand how the fact that one can cut the message into meaningful pieces corresponds in some abstract sense to the structure of the universe. Suppose that (by mutation) you obtain new messages like

$$(\text{sentence } A^*, \text{sentence } B, \text{sentence } C)$$

or

$$(\text{sentence } A, \text{sentence } B^*, \text{sentence } C),$$

and so on. Let us assume that these mutations are not too disastrous, so that the messages $(ABC \ldots)$, $(A^*BC \ldots)$, $(AB^*C \ldots)$ all give a fairly high value to the function E. This does not imply that the recombined message $(A^*B^*C \ldots)$ gives a high value to the function E. Putting together two reasonable mutations might give a catastrophic result, but it often does not. In other words, often (A^*B^*C . . .) is a reasonable genetic message if (A^*BC . . .) and (AB^*C . . .) are, and this expresses at the level of the function E that the universe is not totally meaningless. In fact, the above argument remains true if A, B, C, . . . are pieces of genes or individual letters (= bases) rather than genes.

We have reached an important conceptual conclusion. Let me repeat it. The fact that there is some order in the universe expresses itself at the level of your message of life. It says that it makes sense to recombine mutated messages (A^*BC . . .) and (AB^*C . . .) into a message (A^*B^*C . . .). The process by which recombination is effected is called sexuality.[3] And you, the Creator, seeing that recombination is good for your message, invent sex, and bestow it upon your creatures. This then is the true meaning of sex: that there is some regularity in the universe, and that genetic recombination is therefore useful.

Instead of changing a letter at a time in the genetic message by mutation, there is now the possibility of replacing a word or a sentence by another word or sentence. This is of course much more intelligent. (Note that one can do other things as well, like deleting some parts of the genetic message, or keeping several copies of them.)

With the advent of sex, then, the evolution of life can proceed much faster. Mutations are still occurring, of course, but a more intelligent innovative process is now also at work—the reshuffling of genetic messages. And after the reshuffling, selection operates, of course, to keep the fit and the lucky.[4]

Sex has thus made life much more interesting, and one might easily be carried away, and give a lyrical description of genes collaborating enthusiastically to bring life to higher and higher values of the function E(message).

A more sober picture emerges from modern studies, and it is summarized in the title of a fascinating book, *The Selfish*

Gene,[5] by the British biologist Richard Dawkins. Remember that genes are defined as the elementary meaningful pieces of the genetic messages. In the absence of mutations, they reproduce identical copies of themselves and are thus potentially immortal. The plants or animals are just the mortal vehicles that carry them. There is reason to believe that many genes are hitchhikers on these mortal vehicles and do nothing useful at all (or they may actually be harmful). The cohabitation of many selfish genes is not an easy affair. It is quite wasteful, and we would like to introduce some discipline into the assembly of genes.

What should we do? We turn again to you, the great scientist, the creator of life, the inventor of sex, to give us an idea of how to make your genetic message work more efficiently.

. . . ?

What do you mean, it is all a misunderstanding? You want to take no more responsibility for the creation of life? Or for its evolution? Are you sure?

This is terribly disappointing. You have abandoned your creatures, and now we have to write a new script. To start all over again . . .

So, the stars, the galaxies, and all that, have come into existence. We don't really know how. But there is no serious reason why they shouldn't be there either. The universe has quite a bit of randomness in it, but also quite a bit of structure. And life has come into the universe. Quite easily, it appears,[6] but we don't know exactly how. The little genetic messages that are the essence of life faced the challenge of the randomness of the universe and adapted to it by trial and error. Then, the little genetic messages discovered the art of recombining, which is called sexuality. And it was a good discovery for them, because it gave them a chance to exploit some of the structure of the universe.

The genetic messages of life are assemblies of selfish genes.

But natural selection sees to it that these genes function in a way that is not too wasteful, not too inefficient. And life has created a proliferation of shapes and devices to make use of the world, to take advantage of the regularities of the structure of the universe.

Because there are regularities in the structure of the universe, and because life can take advantage of it, a new feature of life, which we call *intelligence*, has slowly emerged.

CHAPTER 25

Intelligence

David Marr was a specialist in visual information processing and artificial intelligence who worked at the Massachusetts Institute of Technology. His book *Vision*[1] is one of the more important contributions to scientific literature in recent years. David Marr started writing the book when he learned that he had leukemia and did not have much time left to live. *Vision* therefore cuts through the pompous ritualistic nonsense so common in scientific literature and goes straight to the basic questions.

The information that comes to our eyes is processed in various stages from the retina to the visual cortex (a region at the back of the brain). The whole visual processing system performs splendidly to analyze what is going on around us. Some natural questions are: How is our visual system made? Exactly how does it work? How did it come into being? But David Marr asks other questions as well: Suppose we wanted to invent a visual system, starting from scratch, what would be the options? This, if you like, is now an engineering problem. How good is the biological solution to this problem? We know bits and pieces of the answers to all these various questions. Putting these together, we arrive at a grand picture that is very convincing, even if a number of the details are shaky.

For our purposes, the important result is this: Our visual system is constructed to cope with a definite physical reality. That is what comes out loud and clear from David Marr's analysis. Our visual system is not just a general-purpose gadget for analyzing patterns of light intensity and color. It is a device for seeing objects in three-dimensional space, objects that are limited by two-dimensional surfaces, which are in turn limited by edges. The visual system has to see the edges, reconstruct the surfaces, and interpret them in terms of objects lit in a certain way and positioned in a certain manner with respect to the viewer.

When we open our eyes we receive an enormous amount of information from the outside world. But because this outside world has a lot of structure, the messages received by the eyes are highly redundant. The visual system, making assumptions on the class of allowed messages, performs data compression. This data compression begins at the level of the retina, and even before reaching the visual cortex the visual messages are already highly processed and compressed. What we see are interpreted images, interpreted by a visual system that has been shaped by natural evolution to cope with a certain type of outside physical reality.

Let us return to the engineering problem of inventing an efficient visual system. This is a problem in *artificial intelligence*. Why *intelligence*? What we call intelligence is the activity of the mind and takes place in the brain. Intelligence guides our actions on the basis of what we perceive from the outside universe, and the interpretation of visual messages is therefore part of it.

To understand intelligence, a natural idea is to study the brain: investigate its anatomy, use electrodes to analyze its electrical activity, look at its cells under the microscope, and so on. All this has been done, of course, and yields important information (notably on the visual system). Direct investigation of the brain has its limitations, however. It would be hard to reconstruct a natural language like English by looking at a brain. Yet language presumably plays an important role in the organization of human intelligence. As the question of language shows, understanding intelligence is not likely to be an easy problem, and it is unwise to limit oneself to any single methodology, such as neurophysiology or psychology.

An engineer's approach is particularly natural and appropriate for investigating the visual system. An approach of this kind, remarkably, has also been used by Sigmund Freud to analyze sexual instinct. What Freud calls sex is not quite the same thing that we called sex in the last chapter, although the two concepts are related.[2] The Viennese founder of psychoanalysis described a number of *component instincts* (often related to specific erotogenic zones: oral, anal, . . .), and explained sexual instinct in terms of those. The component instincts appear separately in small children. In the natural course of things they later become organized into functional sexual behavior. So-called perversions occur when the component instincts fail to get integrated as they normally should (what is called normal here is that which is favored by natural selection; natural selection will obviously favor the behavior that leads to procreation).

The sexual instinct and the visual system are both understandable on the basis of their function. "Mistakes" of the system, i.e., sexual perversions in one case and visual illusions in the other, guide our interpretation. For the visual system, we have furthermore a rather detailed understanding of how the information is processed from the retina to the brain. The study of the sexual instinct and its component instincts does not benefit from such detailed anatomical and functional studies, and the situation is far worse for other problems that arise in psychoanalysis. In fact, the glory—but also the tragedy—of psychoanalysis lies in its methodological isolation, and this has led to scorn from a number of scientists. Freud himself was a scientist, and he founded psychoanalysis as part of science, but it has been drifting away under his followers. One can only hope that methodological progress will cause a reversal of this tendency. After all, psychoanalysis concerns itself with "brain software" problems, which should at some point make fruitful contact with the "hardware" studies of the neurosciences.

Let us now go back to intelligence. By putting together a sex instinct, a visual system, and a few other such devices, one can probably get a reasonable brain for a rat or for a monkey. But isn't the human intellect something totally different and incomparably superior? Well, maybe not. One reason to think that the difference is not so extreme is that from the point of view of evolution, the differentiation of the human brain has taken relatively little time (a few million years, and probably much less for the development of complex natural languages). The extra development needed to obtain the human brain from that of a rat or a monkey is thus probably just "the icing on the cake" as far as new gadgetry is concerned. In other words, the specifically human abilities of using tools and learning complex languages have probably been easy developments, even if they have gone with a considerable increase in brain size.

Of course, we have intellectual possibilities vastly superior to those of rats and monkeys: we can argue about the theological problem of predestination, read and enjoy poetry, and prove that the sequence of prime numbers is infinite. But the brain that we use has basically the same gadgetry as that of a rat or monkey. It is pathetic that this superior brain of ours has difficulty with simple arithmetic operations, does not keep time correctly, and cannot easily memorize a few thousand digits. (This is why we use calculators, clocks, calendars, and directories.) In the typically "superior" activity of doing science we seem to use mostly our speech system and our visual system. Getting the visual system involved is a great asset, and this is why geometrization of mathematics is important.

Let us try to summarize. Our brain and intelligence have a basis consisting of gadgetry strictly geared to surviving in a certain type of environment. Evolution has, rather recently, added to these basic brain skills some higher functions that perform very flexibly. Possession of these higher functions was of course beneficial, and encouraged by natural evolution. As a byproduct these higher functions have also allowed humans to develop scientific knowledge. But this, it seems to me, was an accident. The human brain lacks some basic functions that are desirable for doing science, like the ability to compute quickly and reliably, or the ability to store large amounts of data. In spite of these shortcomings, human science has developed, and we are thus able to understand a lot more about the nature of things than we had any right to hope for.

We live, apparently, in a world full of three-dimensional objects limited by two-dimensional surfaces.[3] Therefore it is not astonishing that our brain can cope with such objects: this skill is useful for survival and encouraged by natural selection. But natural selection does not explain how we came to understand the chemistry of stars, or subtle properties of prime numbers. Natural selection explains only that humans have acquired higher intellectual functions; it cannot explain why so much is understandable about the physical universe, or the abstract world of mathematics.

We have argued that the physical universe should exhibit a lot of randomness. We have argued that many mathematical assertions should be unprovable. Yet, remarkably, we understand a lot about the physical universe and about mathematics.

What we call understanding is very much linked to the specific nature of human intelligence. For instance, we make heavy use of natural languages in mathematics, because our brain can't cope with completely formalized mathematical languages, which in principle would be much better. (Mathematical literature looks formal and incomprehensible enough, but is not what mathematicians call formalized mathematical language; call it semiformal if you like.) We present our mathematical knowledge in the form of brief theorems because we couldn't stomach really long formulations. There is no doubt that nonhuman intelligent beings would do mathematics rather differently than we do, and we get a glimpse of that from the growing use of computers as aids in mathematical investigations. (Present-day computers can't cope with natural languages, but are not averse to the use of very long codes.) In brief, the way we do mathematics is human, very much so. But mathematicians have no doubt that there is a mathematical reality beyond our puny existence. We discover mathematical truth, we do not create it. We ask ourselves what seems to be a natural question and start working on it, and not uncommonly we find the solution (or someone else does). And we know that the answer could not have been different. And the strange thing is that, because of Gödel's theorem, we had no guarantee that our question could be solved. We don't understand why the world of mathematical truth is accessible to us. Yet, wonderfully, it is . . .

The comprehensibility of the physical universe in terms of mathematical structures is no less amazing. The Hungarian-American physicist Eugene Wigner has described his amazement in a paper with a very expressive title: *The unreasonable effectiveness of mathematics in the natural sciences*.[4] We have learned how vast the universe is, and how insignificant we are in it. Yet, wonderfully, we can probe into the depth of this universe, and understand it.

CHAPTER 26

Epilogue: Science

Let us jump a few thousand years back in time. Night is falling, the day's work is over, and the oil lamps are lit. We talk about local events, and how we should plan rural activities by looking at the constellations in the sky. We discuss the tales of travelers and the strange languages that they speak. There is an argument about the attributes of the gods, or a point of law, or the medicinal virtues of some plant. Intellectual curiosity is there, the urge to understand the secrets of the vast world and the nature of things. And we turn this curiosity to all kinds of problems: how to interpret dreams and learn about the future, how to understand the signs in the sky, or how to make a right angle with a piece of string (make a triangle with sides 3, 4, and 5).

And now, a few thousand years later, as we look back, we see that some of the topics of the old discussions have been forgotten: the attributes of ancient gods no longer interest us too much. Some questions have not changed greatly: what is the true nature of art? and what is consciousness? But the study of yet other problems has led to the tremendous accomplishments of science and technology, which have completely changed our human condition. From making right angles with a piece of string, mathematics has developed. Trying to understand the motion of stellar bodies has led to the creation of mechanics and physics. And, later, biology and modern medicine have arisen, replacing the study of medicinal plants.

Science has fared differently than other areas of human curiosity, not because the curiosity was different, but because the objects and the concepts put forward were different. It has been more profitable to argue about the properties of triangles than about the interpretation of dreams. It has been more rewarding to study the motion of the pendulum than the nature of consciousness. Sometimes the old philosophical problems are clarified by science; sometimes they subvert science. But the questions that are suggested by introspection often remain unanswered, and when the answers come they tend to be intellectually convincing rather than psychologically satisfying.[1]

Chance and *randomness* did not look like very promising topics for precise investigation, and were in fact shunned by many early scientists. Yet they play now a central role in our understanding of the nature of things. It was the purpose of the present book to give an idea of this role. We have seen how we idealize the world around us in physical theories, and how *chaos* limits the intellectual control that we have on the evolution of the world. We have seen how a correct assessment of chance and *predictability* is important for everyday life and for history. We have introduced *entropy*, which measures the amount of randomness in the molecular chaos of a liter of water. We have had a glimpse of *complexity* problems, and we have seen how useful information may be very hard to reach. And we have found chance even in the properties of the natural numbers 1, 2, 3, . . .

Now, let us have a last look at the people who do science.

From discussions with a number of colleagues, I have come to the conclusion that the physicists of my generation fall into two broad classes. Some developed their scientific tastes doing amusing chemistry when they were young. Others were more attracted by electricity and mechanics, and spent time dismantling radio sets, alarm clocks, and the like. I was a resolute chemist, and I occasionally have a good time comparing memories, with some colleague or other, of the various crazy things we did in the old days. Like preparing nitroglycerine or mercury fulminate, or boiling concentrated sulfuric acid in a Pyrex test tube. (I do not really advocate doing any of these things, especially the latter.) When I asked the American physicist John Wheeler if he belonged to the chemical or electro-mechanical category, he said, ''Both.'' And his wife, who was present, seized his hand, saying, ''Show your little finger, Johnny.'' And Johnny had to show a finger with a piece missing as a result of doing some ''amusing experiment'' in his youth. The physicist Murray Gell'Mann told me, however, that he did not indulge in amusing science, but read a lot of science fiction instead.

Because of the problems of drugs and terrorism, amusing chemistry tends to be discouraged these days, and there is also less and less fun to be had dismantling radio sets or alarm clocks (there is so little left inside to be seen). Therefore, people get their kicks playing with computers, and this must produce a different kind of physicist. In all cases, however, the career of a physicist begins with some kind of fascination—of a magical kind, perhaps, in the case of amusing chemistry; of a more logical kind in the case of electro-mechanical devices and computers. And I dismiss as irrelevant the case of people who make a living by ''doing research'' but who would rather watch baseball on TV when they have a choice.

Mathematicians, like physicists, are pushed by a strong fascination. Research in mathematics is hard, it is intellectually painful even if it is rewarding, and you wouldn't do it without some strong urge.

What is the origin of the urge, the fascination that drives physicists, mathematicians, and presumably other scientists as well? Psychoanalysis suggests that it is sexual curiosity. You start by asking where little babies come from, one thing leads to another, and you find yourself preparing nitroglycerine or solving differential equations. This explanation is somewhat irritating, and therefore probably basically correct. Sexual curiosity is at the root of science, but it is relayed by something else, namely the fact that the world is understandable. A purely psychological approach to science would miss the importance of the comprehensibility of mathematics, and of "the unreasonable effectiveness of mathematics in the natural sciences." In fact, some scientists in the "soft" sciences seem to miss this as well. But mathematicians and physicists know that they deal with a reality that has laws of its own, a reality above our little psychological problems, a reality that is strange, fascinating, and in some sense beautiful.

At this point I would have wanted to write a moving description of the greatness of answering the riddles of science. But I see that you won't let me . . . You want to talk about Oedipus, who so smugly answered the riddle of the Sphinx and started in this manner a chain of events so catastrophic, so disastrous, that it has kept dramatic authors and psychoanalysts busy for the next three thousand years. Scientists too begin by answering riddles, then blow up pieces of fingers, and then perhaps the entire planet. Should not science behave more responsibly?

The answer to that last question is clear: science is totally amoral, and completely irresponsible. Individual scientists act according to their own individual sense of moral responsibility (or lack of it), but they act as humans, not as representatives of Science. Let us take an example. What we used to call *Nature* has been downgraded to become *our environment*, and is being further degraded to become our junkyard. Is this the fault of science? Science can indeed help in destroying Nature, but it can also help in protecting the environment, or it can help in assessing pollution: the decisions are all human. Science answers questions, at least sometimes, but it does not make decisions. Humans make decisions, or at least sometimes they do.

It is difficult to assess what options really are open to mankind. Is doom imminent? Or can it be indefinitely postponed? The brain that we use is the same as that of our Stone Age ancestors, and has shown amazing flexibility. Instead of running on foot and hunting with spears, modern humans drive cars and sell insurance. And unless some cataclysm occurs rather soon, there will be more changes, more progress. For serious technical work, at least, our obsolescent Stone Age brains will be progressively replaced by faster, more powerful, more reliable machines. And Science will improve on our antiquated genetic copying mechanisms, avoiding all kinds of terrible diseases. And we can't say *No*. For sociological reasons we don't have the option of saying that we refuse all these beautiful improvements. But will mankind be able to survive the changes that we cannot avoid making to our physical and cultural environment? We do not know.

Now, as before, the human future remains inscrutable, and we do not know if we are heading for a nobler future or toward unavoidable self-destruction.

Notes

Chapter 1. Chance

1. *The four-color theorem.* Suppose that we have a geographical map on a sphere or in the plane. The map shows various countries, and for simplicity we assume that there are no seas. Also, each country is *connected* (not composed of disjoint pieces). We want to give a color to each country so that two countries with a common border have different colors. (We accept the same color for two countries that have only a finite number of common boundary points.) How many colors do we need? Answer: four colors suffice in all cases. This is the four-color theorem.

The solution of the four-color problem is due to Kenneth Appel and Wolfgang Haken. The technical papers are: K. Appel and W. Haken, "Every planar map is four colorable, Part I: Discharging," *Illinois J. Math.* 21 (1977): 429–90; K. Appel, W. Haken, and J. Koch, "Every planar map is four colorable, Part II: Reducibility," *Illinois J. Math.* 21 (1977): 491–567.

For more popular expositions, see K. Appel and W. Haken, "The solution of the four-color-map problem," *Scientific American*, October 1977, pp. 108–21; K. Appel and W. Haken, "The four color proof suffices," *The Mathematical Intelligencer* 8 (1986): 10–20.

2. For a brief introduction to the problem of listing the simple finite groups, see J. H. Conway, "Monsters and Moonshine," *The Mathematical Intelligencer* 2 (1980): 165–71. It should be mentioned that the classification of simple finite groups has involved a lot of computer work as well as a huge amount of mathematicians' time.

3. The standard biography of Newton is R. Westfall's *Never at Rest* (Cambridge: Cambridge University Press, 1980). The interplay between Newton's various intellectual interests is fascinating. These interests range from the greatest achievements in mathematics and physics to disreputable speculations (by present-day standards) about alchemy, history, and religion. It is tempting to apply censorship to Newton's intellectual production and decree that some is good and the rest better forgotten. If, however, we want to understand the process of intellectual creation in

Newton's mind, we cannot forget his disreputable speculations. In his desire to grasp the meaning of the universe the research on the prophecies or alchemy was not less important than the work on gravitation or differential calculus. A lot, obviously, remains to be understood on how Newton's mind functioned. One unfortunate fact seems to emerge from Westfall's book: the great Newton apparently had no sense of humor of any recognizable kind.

Chapter 2. Mathematics and Physics

1. Mathematicians are in fact a somewhat heterogeneous group. This makes sense: Some mathematicians approach problems in a straightforward way, and owe their success to their great technical power. Others turn around a problem until they find a subtle trick that allows an easy solution. (Note that there is not always such a subtle trick.) Not all mathematicians, therefore, are alike, and some don't even look like mathematicians. But often there is an air of family between mathematicians, or more generally between professional scientists. Even physically. I have more than once found my way to a scientific meeting in an unfamiliar place by following in the street a person who looked like a colleague. Other people have made the same observation.

2. See Chapter 22 and Chapter 23. In brief, Gödel's incompleteness theorem is as follows. Within the framework of generally accepted basic assertions concerning the integers 1, 2, 3, . . . , Gödel shows that some assertions can be neither proved nor disproved: these are *undecidable* assertions. If one increases the number of basic assertions, there will nevertheless always remain some undecidable assertions.

3. See H. Poincaré, "L'invention mathématique" (Mathematical creation), Chapter 3 in *Science et Méthode* (Paris: Ernest Flammarion, 1908); English translation: *Science and Method* (New York: Dover, 1952). See also J. Hadamard, *The Psychology of Invention in the Mathematical Field* (Princeton: Princeton University Press, 1945; reprint, enlarged edition, New York: Dover, 1949).

Poincaré discusses the example of a problem about which he was no longer thinking consciously, and for which the solution came to him later, abruptly and with complete clarity. Obviously, some unconscious work had been performed. This work would involve what Freud calls *preconscious* rather than the deep unconscious, but putting on a label like preconscious does not quite explain what is going on. The role of the unconscious, or preconscious, in scientific discovery is familiar, I think, to many scientists, but a real understanding of it is missing.

4. Here is an excerpt from Galileo Galilei's *Saggiatore* (of 1623): "Philosophy is written in this very great book which is continuously open in front of our eyes (I mean the universe), but it cannot be understood if one does not first learn the language, and know the characters in which it is written. It is written in mathematical language, and the characters are triangles, circles, and other geometrical figures . . ."

5. The mathematics of a physical theory may go well beyond the operationally defined quantities and introduce objects that are not directly observable, even in principle. The introduction of unobservable objects is of course a very delicate matter, and one may be tempted to refuse it on philosophical grounds. But such a philosophical *a priori* attitude turns out, in some cases at least, to be a bad idea. For instance, the physicist Geoffrey Chew proposed in the late 1950s that particle physicists should concentrate their efforts on the study of a mathematical object called the *S-matrix* (which is closely related to experimental quantities) and disregard unobservable *quantum fields*. Chew's idea was in some sense very reasonable. As it happens, however, the consideration of fields has been extremely fruitful (before and after Chew's proposal), and we wouldn't want to be without them.

Chapter 3. Probabilities

1. The mathematical foundations of the calculus of probabilities were made mathematically respectable by Kolmogorov (the same man whose theory on the psychology of mathematicians we discussed at the beginning of Chapter 2, and whose theory of turbulence will be mentioned later). The standard reference is A. N. Kolmogorov, *Grundbegriffe der Wahrscheinlichkeitsrechnung*, Erg. Math. (Berlin: Springer, 1933); English translation: *Foundations of the Theory of Probability* (New York: Chelsea, 1950).

2. We insist on giving a physical definition of the independence of events. Saying that two events are independent when they have "nothing to do with each other" can hardly be called an operational definition, though. It would be better to say that it is a metaphysical principle that suggests operational definitions in specific cases, and the validity of these operational definitions can then be checked from the consequences. But why not rather use the mathematical definition of independence [i.e., basically assertion (3)] and verify it by statistical tests? This is a neat way to present things in principle, and it is the way used in textbooks, but *not* the way used in practice.

In fact, statistical tests are a heavy and often unconvincing machinery.

So, scientists first *guess* that two events are independent because they have nothing to do with each other. Then they will think of possible reasons why the independence would be spoiled. And only as a last resort will they use statistical testing.

CHAPTER 4. LOTTERIES AND HOROSCOPES

1. Actually, buying a lottery ticket once in a while (or gambling small sums) may be reasonable if you get adequate amusement out of it. Economics textbooks discuss the logic of this, and also the related problem of insurance (why it makes sense to buy insurance even though you know that the insurance company makes an unfair profit on you). What we have shown is that buying many lottery tickets in the hope of getting rich is not a good idea.

2. In a large number N of trials, let $N(A)$ be the number of those when the event "A" is realized, and $N(A$ and B) the number of those in which "A" and "B" are realized. The probability of "B" knowing that "A" is realized should be approximately

$$\frac{N(A \text{ and } B)}{N(A)},$$

which is equal to

$$\frac{N(A \text{ and } B)}{N} \div \frac{N(A)}{N}$$

and therefore approximately equal to

$$\text{proba(``}A \text{ and } B\text{'')} \div \text{proba(``}A\text{'')}.$$

It is therefore reasonable to make the *definition*

$$\text{proba(``}B\text{,'' knowing that ``}A\text{'' is realized)} = \frac{\text{proba(``}A \text{ and } B\text{'')}}{\text{proba(``}A\text{'')}}.$$

(This is a so-called *conditional probability*.) If "A" and "B" are independent, (3) implies that the right-hand side is

$$\frac{\text{proba(``}A\text{'')} \times \text{proba(``}B\text{'')}}{\text{proba(``}A\text{'')}} = \text{proba(``}B\text{''}),$$

and this proves (4).

3. Let me give here a brief technical discussion of how it is that the weather can depend sensitively on the position of Venus a few weeks ago, and be statistically independent from it. Let x be some initial state of the system under consideration, i.e., the universe, or—better—an idealization of the universe, describing among other things the position of Venus and the weather where you are. If the initial state x refers to the situation a few weeks ago, the situation this afternoon will be a state $f^t x$; here f^t is called the *time evolution operator* and is a transformation of the space of states of our system (corresponding to the time evolution from a few weeks ago to this afternoon). There is a set A of possible initial conditions for our system, between which we cannot distinguish: this expresses the fact that we do not know the initial condition with complete precision. (For the sake of argument, we may assume that only the initial position of Venus is not known with complete precision.) The different possibilities for the weather this afternoon are described by all the points in the set $f^t A$. And, because of the phenomenon of sensitive dependence on initial condition, to be discussed in later chapters, the set $f^t A$ will no longer be small but will in fact cover all sorts of different possibilities for the weather. Let now B be the set of states describing rain this afternoon. Part of $f^t A$ will be in B, part will be outside, and the effect of Venus a few weeks ago thus prevents us from saying whether or not you will have rain this afternoon. The states of the universe having rain this afternoon (where you are) and compatible with what we know of the situation a few weeks ago are the points of the intersection $(f^t A) \cap B$. Can we say something about this intersection?

To be able to progress in the discussion we make use of the fact that for many time evolutions there is a natural *probability measure* m that does not change under the time evolution, and describes the probability of various events. For instance, $m(f^t A) = m(A)$ is the probability of the event "A" associated with our initial condition. Furthermore, $m((f^t A) \cap B)$ is the probability of the event "A" a few weeks ago and "B" this afternoon. It happens that in many cases, for large t,

$$m((f^t A) \cap B) \approx m(A) \times m(B).$$

This property, called *mixing*, means that the set $f^t A$ is so convoluted that the fraction of it in B is proportional to the size of B [measured by $m(B)$].

If we interpret the above property of mixing in terms of probabilities, we see that it gives precisely the same result as assuming that rain this afternoon and the position of Venus a few weeks ago are (statistically) independent. [The fact that $m(A) = 0$ is a minor technical difficulty, handled by taking a suitable limit.]

The above justification of statistical independence is of course unsatisfactory for a mathematician, who would demand a *proof*. And we are very far from being able to supply one: the problem is just too hard. If you are a physicist, you will not be put off by the absence of mathematical proof, but you will ask for other things. You will first demand evidence of sensitive dependence on initial condition in our problem, and will like to know how many weeks is "a few weeks" (this will be discussed in subsequent chapters). You will then want to define precisely what is meant by the position of Venus (if you are careless, the position of Venus will be correlated with the time of the year, and therefore with the seasonal weather). You will also look into the problem of mixing. And since this is very difficult to attack directly, you will try to see what could go wrong with the assumptions of statistical independence of rain and the position of Venus. One thing that might go wrong is an intelligent agent modifying the weather in accordance with observations of Venus. But with present technology this is unlikely. Finally, if the matter is of sufficient interest, you will embark on a series of observations and statistical tests of independence of the weather and the position of Venus.

Our discussion leaves at least one question open: what do we mean by an *intelligent agent*? All we can say at this point is that an intelligent agent introduces correlations where you wouldn't otherwise expect any. If you think about it, this is not a bad characterization of intelligence.

Chapter 5. Classical Determinism

1. *Newton's equation*: Consider N points with masses $m_1, \ldots, m_N$ (positive numbers) and positions $x_1, \ldots, x_N$ (vectors in 3-space); then Newton's equation is

$$m_i \frac{d^2}{dt^2} x_i = F_i \qquad \text{for } i = 1, \ldots, N,$$

where F_i is the force on the ith particle (a 3-vector). We speak of Newton's equation in the singular, but there are actually $3N$ equations, because each position has 3 components. The *gravitational force* is given by

$$F_i = \gamma \sum_{j \neq i} m_i m_j \frac{x_j - x_i}{|x_j - x_i|^3},$$

where γ is the *constant of gravitation*. This is the force used, for instance, to study the motion of the planets around the sun. If the positions x_i and the velocities dx_i/dt are known at some initial time, it is in principle possible to deduce them at a different time, from Newton's equation. I said *in principle* because the existence and uniqueness of solutions of Newton's equation with gravitational forces is not guaranteed for all initial conditions. Also, when N is equal to 3 or more the solutions cannot be obtained in explicit analytical form, and their study becomes very delicate.

2. P. S. Laplace, *Essai philosophique sur les probabilités* (Paris: Courcier, 1814).

3. R. Thom, "Halte au hasard, silence au bruit (, mort aux parasites)"; Edgar Morin, "Au-delà du déterminisme: Le dialogue de l'ordre et du désordre"; Ilya Prigogine, "Loi, histoire . . . et désertion." These articles were published in the French journal *Le Débat* in 1980 (no. 3 and no. 6), and Thom omitted the "mort aux parasites" in the printed version. These and other articles are now reprinted in the collective work *La querelle du déterminisme: Philosophie de la science d'aujourd'hui* (Paris: Gallimard, 1990).

4. E. Schrödinger, "Indeterminism and free will," *Nature*, July 4, 1936, pp. 13–14. This paper is reprinted in E. Schrödinger, *Gesammelte Abhandlungen* (Vienna: Vieweg, 1984), vol. 4, pp. 364–65.

Chapter 6. Games

1. *The minimax theorem*. We consider a *finite zero-sum two-person game*. There are thus two players, A and B. Player A can choose among M options (these are labeled $1, \ldots, M$), and player B has N options (these are labeled $1, \ldots, N$). That the game is *finite* means that M and N are finite. Choice i of player A and j of player B produce a payoff K_{ij} for player A and $-K_{ij}$ for player B. That the game is *zero-sum* means that the amount $|K_{ij}|$ gained by one player is lost by the other. Suppose now that player A selects his options with probabilities $p_1, \ldots, p_M$ and player B her options with probabilities $q_1, \ldots, q_N$. The average payoff to player A is then

$$\sum_{i=1}^{M} \sum_{j=1}^{N} K_{ij}\, p_i\, q_j,$$

and it is minus that amount for player B. Player A will try to make his average payoff as large as possible for the worst possible choice of q's by player B. This gives

$$\min_{(q_1, \ldots, q_N)} \max_{(p_1, \ldots, p_M)} \sum_i \sum_j K_{ij}\, p_i q_j. \tag{1}$$

The corresponding amount for player B is

$$\min_{(p, \ldots, p_M)} \max_{(q_1, \ldots, q_N)} \sum_i \sum_j (-K_{ij})\, p_i q_j$$

$$= - \max_{(p_1, \ldots, p_M)} \min_{(q_1, \ldots, q_N)} \sum_i \sum_j K_{ij}\, p_i q_j. \qquad (2)$$

The minimax theorem states that (2) is minus (1), i.e.,

$$\min \max \sum_i \sum_j K_{ij} p_i q_j = \max \min \sum_i \sum_j K_{ij} p_i q_j, \qquad (3)$$

where the min and max are conditional on $p_1, \ldots, p_M, q_1, \ldots, q_N \geq 0$ and $\sum p_i = \sum q_j = 1$.

Note that if players A and B did not use probabilistic strategies, but stuck instead to *pure* strategies, they would not have a minimax theorem, because in general

$$\min_j \max_i K_{ij} \neq \max_i \min_j K_{ij}.$$

What happens, however, in this situation is that one of the players finds it advantageous to turn to a probabilistic strategy.

This minimax theorem is due to John von Neumann (J. v. Neumann and O. Morgenstern, *Theory of Games and Economic Behavior* [Princeton: Princeton University Press, 1944]).

How do we obtain the value K of the minimax (3) and the p_i, q_j, which give the optimal strategies for players A and B? These quantities are determined by the *linear* conditions

$$p_i \geq 0, \sum_j K_{ij} q_j \leq K \quad \text{for} \quad i = 1, \ldots, M$$

$$q_j \geq 0, \sum_i K_{ij} q_i \geq K \quad \text{for} \quad j = 1, \ldots, N$$

$$\sum_i p_i = \sum_j q_j = 1.$$

Finding a solution of such a system of linear equalities and inequalities is a problem of *linear programming*.

For the specific case of the table of payoffs shown in the text one finds $p_1 = 0, p_2 = 0.45, p_3 = 0.55, q_1 = 0.6, q_2 = 0.4, q_3 = q_4 = 0, K = 3.4$.

Chapter 7. Sensitive Dependence on Initial Condition

1. The growth (= time derivative) of the distance between the real and imaginary balls is proportional to the angle between the trajectories.

Therefore the distance between the two balls is estimated by the integral of an exponential, and this is again (up to an additive constant) an exponential:

$$\int_0^t A\, e^{\alpha s}\, ds = \frac{A}{\alpha}\,(e^{\alpha t} - 1).$$

Of course, the assumption of one shock per second is an approximation, and even so the increase of the angle is only roughly exponential. But the only serious difficulty with our argument is that it holds only for a *small* distance between the balls.

2. Ya. G. Sinai, "Dynamical systems with elastic reflections," *Uspekhi Mat. Nauk* 25, no. 2 (1970): 137–92; English translation: *Russian Math. Surveys* 25, no. 2 (1970): 137–89. This is the original publication (quite technical); it has been followed by a number of other papers by various authors.

Chapter 8. Hadamard, Duhem, and Poincaré

1. J. Hadamard, "Les surfaces à courbures opposées et leurs lignes géodésiques," *J. Math. pures et appl.* 4 (1898): 27–73, reprinted in *Oeuvres de Jacques Hadamard* (Paris: CNRS, 1968), vol. 2, pp. 729–75. Note that Hadamard's original paper already explicitly contains the remark that if there is any error in the initial condition, the long-term behavior of the system is unpredictable.

2. It is easiest to study compact surfaces of *constant* negative curvature. (Their only drawback is that, unlike Hadamard's surface, they cannot be realized in three-dimensional Euclidean space.) You probably remember Euclid's postulate that through a point outside a straight line there passes one and only one parallel to that straight line. And you know also that *non-Euclidean* geometries can be constructed in which this postulate does not hold. In particular, in the Lobachevsky plane there are many parallels to a given line passing through a point outside of it. Therefore, in the Lobachevsky plane, two points moving straight on parallel lines usually move away from each other! The billiards with constant negative curvature is obtained by cutting a piece of Lobachevsky plane and gluing the edges together to form a smooth closed surface (that this can be done requires of course a proof). It is then not too hard to believe that on such a billiard table the straight-line motion exhibits sensitive dependence on initial condition.

3. In French, the title of the section is "Exemple de déduction mathématique à tout jamais inutilisable," in P. Duhem, *La théorie physique: Son objet et sa structure* (Paris: Chevalier et Rivière, 1906). This reference was pointed out to me by René Thom.

4. H. Poincaré, *Science et Méthode* (see note 3 of Chapter 2). The relevant chapter is Chapter 4, "Le hasard" (Chance).

5. Small causes can have large effects even when there is no sensitive dependence on initial condition. The large effect may come from waiting a very long time, as Poincaré observes.

Another interesting case is that of systems with several equilibrium states; it may be very difficult to determine which initial conditions will eventually lead to one or another equilibrium state. This happens when the boundaries of the *basins of attraction* of the various equilibrium states are very convoluted, as is frequently the case. A simple example is provided by a *magnetic pendulum*—a little magnet suspended by a rigid rod over several other magnets. If one kicks such a pendulum, it starts oscillating in a complicated way, and it is hard to guess in what equilibrium position it will end up (there are usually several such equilibria). For pictures of convoluted boundaries, see for instance S. McDonald, C. Grebogi, E. Ott, and J. Yorke, "Fractal basin boundaries," *Physica* 17D (1985): 125–53.

Poincaré also observes that what we call chance may arise from our lack of muscular control, and gives the example of the roulette game. Coin tossing is similar, and some trained people are able to toss a coin with completely predictable result.

Chapter 9. Turbulence: Modes

1. "The little 'gravitic' Doctor": I owe the story to George Uhlenbeck, and other facts about Th. De Donder to Marcel Demeur.

2. A lot of data about the fascination that is at the root of scientific work could be gathered by interviewing scientists. The interpretation of the data would be delicate but might give new insight into the psychology of scientific discovery. Scientists gone insane or senile would be of particular interest in such a study, because of the greater transparency of their motivations. (Many people unfortunately lose interest in science at an early age, and remain otherwise completely normal. I have known, however, at least one fantastic example of a great physicist whose judgment was rather diminished for matters of everyday life, but who appeared great and lucid again when he spoke about science.)

3. J. Leray, "Sur le mouvement d'un liquide visqueux emplissant l'espace," *Acta Math.* 63 (1934): 193–248.

4. H. Poincaré, *Théorie des tourbillons* (Paris: Carré et Naud, 1892).

5. P. Cvitanović, *Universality in Chaos* (Bristol: Adam Hilger, 1984); Hao Bai-Lin, *Chaos* (Singapore: World Scientific, 1984). For a popular presentation see J. Gleick, *Chaos* (New York: Viking, 1987). This is a very readable journalistic account, but not always to be trusted on matters of historical accuracy or scientific priority. An excellent introductory book is I. Steward, *Does God Play Dice? The New Mathematics of Chaos* (London: Penguin, 1990).

6. The original publications are: L. D. Landau, "On the problem of turbulence," *Dokl. Akad. Nauk SSSR* 44, no. 8 (1944): 339–42; E. Hopf, "A mathematical example displaying the features of turbulence," *Commun. Pure Appl. Math.* 1 (1948): 303–22. Landau's ideas are accessible in English in § 27 of L. D. Landau and E. M. Lifshitz, *Fluid Mechanics* (Oxford: Pergamon Press, 1959).

7. T. S. Kuhn, *The Structure of Scientific Revolutions*, 2d ed. (Chicago: University of Chicago Press, 1970). I am not an uncritical believer in Kuhn's ideas; in particular, they appear to me of little relevance to pure mathematics. The physical concepts of *modes* and *chaos* seem, however, to fit rather well Kuhn's description of *paradigms*.

8. S. Smale, "Differentiable dynamical systems," *Bull. Amer. Math. Soc.* 73 (1967): 747–817.

Chapter 10. Turbulence: Strange Attractors

1. Using the notation of note 3 of Chapter 4, the initial condition x gives after time t a point $f^t x$. If x is replaced by $x + \delta x$, then $f^t x$ is replaced by $f^t x + \delta f^t x$, and if $\delta f^t x = \dfrac{\partial f^t x}{\partial x} \cdot \delta x$ grows exponentially with t, we say that we have sensitive dependence on initial condition. More precisely, we have sensitive dependence on initial condition if the matrix of partial derivatives $\partial f^t x / \partial x$ has norm growing exponentially with t. Consider now a motion described by k angles with initial values $\theta_1, \ldots, \theta_k$, and becoming after time t: $\theta_1 + \omega_1 t, \ldots, \theta_k + \omega_k t \pmod{2\pi}$. Writing

$$f^t(\theta_1, \ldots, \theta_k) = (\theta_1 + \omega_1 t, \ldots, \theta_k + \omega_k t), \qquad (1)$$

we find

$$\delta f^t(\theta_1, \ldots, \theta_k) = (\delta\theta_1, \ldots, \delta\theta_k).$$

The right-hand side is independent of t, and we have therefore no sensitive dependence on initial condition. The time evolutions that can be put in the form (1) by a change of variables are called *quasiperiodic* and have again no sensitive dependence on initial condition. Note that a change of variables as discussed is a parametrization by k angles, corresponding to the superposition of k *modes*. A set that can be parametrized by k angles is a k-*torus* or k-dimensional torus (i.e., a product of k circles).

2. E. N. Lorenz, "Deterministic nonperiodic flow," *J. Atmos. Sci.* 20 (1963): 130–41.

3. D. Ruelle and F. Takens, "On the nature of turbulence," *Commun. Math. Phys.* 20 (1971): 167–192; 23 (1971): 343–44.

4. See note 1, Chapter 10, above.

5. B. Mandelbrot, *Les objets fractals* (Paris: Flammarion, 1975); English version: *The Fractal Geometry of Nature* (San Francisco: Freeman, 1977). Mandelbrot has attracted the attention of scientists on the ubiquity of fractal shapes among natural objects. This was an important and fruitful contribution. What is still missing in general is an understanding of how fractal shapes arise.

Chapter 11. Chaos: A New Paradigm

1. J. B. Mc Laughlin and P. C. Martin, "Transition to turbulence of a statically stressed fluid," *Phys. Rev. Lett.* 33 (1974): 1189–92; J. P. Gollub and H. L. Swinney, "Onset of turbulence in a rotating fluid," *Phys. Rev. Lett.* 35 (1975): 927–30.

2. T. Li and J. A. Yorke, "Period three implies chaos," *Amer. Math. Monthly* 82 (1975): 985–92. In this pleasantly written paper it is shown that, for a large class of maps of a line interval into itself, the existence of a periodic point of period 3 implies the existence of periodic points of every other period. This complicated situation is what is called *chaos* in the paper. The name was remarkably successful, but applied to a different situation! (A time evolution with many periodic orbits often does not show sensitive dependence on initial condition. In fact, the many periodic orbits need not be on an attractor, so that their presence is not relevant to the long-term time evolution of the system.) It was found after a while that the result of Li and Yorke was a special case of an earlier theorem by Šarkovskii. Consider a *unimodal map* f: $[-1,1] \to [-1,1]$, i.e., a continuous map such that $f(-1) = f(1) = -1$, and f is increasing on $[-1,0]$, decreasing on $[0,1]$. Consider now the following unusual order of the positive integers:

$$3 \gtrless 5 \gtrless 7 \gtrless \ldots \gtrless 2{\cdot}3 \gtrless 2{\cdot}5 \gtrless 2{\cdot}7 \gtrless \ldots$$
$$2^n{\cdot}3 \gtrless 2^n{\cdot}5 \gtrless 2^n{\cdot}7 \gtrless \ldots$$
$$2^n \gtrless \ldots 4 \gtrless 2 \gtrless 1$$

(first the odd numbers, then the odd numbers multiplied by 2, 4, 8, . . . , and finally the powers of 2 in decreasing order). Šarkovskii's remarkable theorem is that if $p \gtrless q$ and f has a periodic point of order p (i.e., $f^p x = x$ and $f^m x \neq x$ for $m < p$), then f has a periodic point of period q. In particular, for $p = 3$, we recover the result of Li and Yorke. The original reference is A. N. Šarkovskii, "Coexistence of cycles of a continuous map of a line into itself," *Ukr. Mat. Z.* 16 (1964): 61–71; there are some surprisingly good papers in Ukrainian mathematical journals!

3. M. J. Feigenbaum, "Quantitative universality for a class of nonlinear transformations," *J. Statist. Phys.* 19 (1978): 25–52; also "The universal metric properties of nonlinear transformations," *J. Statist. Phys.* 21 (1979): 669–706. The idea of a rigorous computer-assisted proof is in O. E. Lanford, "A computer-assisted proof of the Feigenbaum conjectures," *Bull. Amer. Math. Soc.* 6 (1982): 427–34; The important transition from one to several dimensions is in P. Collet, J.-P. Eckmann, and H. Koch, "Period doubling bifurcations for families of maps on $\mathbf{R}^n$," *J. Statist. Phys.* 25 (1981): 1–14. This then allows one to deal with continuous-time systems, as occur in physical applications. Incidentally, the word *universality* in Feigenbaum's title refers to a technical feature of the Renormalization Group approach. It does not mean that chaos is always reached via Feigenbaum's period-doubling cascade (which is in fact not particularly frequent). There are many different *routes to chaos*; some particularly important ones are reviewed in J.-P. Eckmann, "Roads to turbulence in dissipative dynamical systems," *Rev. Mod. Phys.* 53 (1981): 643–54. Periodic doubling cascades have been observed in a number of experiments, notably by Albert Libchaber in convection studies; see A. Libchaber, C. Laroche, and S. Fauve, "Period doubling cascade in mercury, a quantitative measurement," *J. de Physique—Lettres* 43L (1982): 211–16.

4. K. Pye and B. Chance, "Sustained sinusoidal oscillations of reduced pyridine nucleotide in a cell-free extract of Saccharomyces carlsbergensis," *Proc. Nat. Acad. Sci. U.S.* 55 (1966): 888–94.

5. D. Ruelle, "Some comments on chemical oscillations," *Trans. N.Y. Acad. Sci. Ser. II* 35 (1973): 66–71.

A few words about rejected papers may be appropriate here. A prerequisite for a successful professional career, for many people, is to have published scientific papers in refereed journals. In other words, appointments and promotions are decided on the basis of number of published papers. This situation forces many individuals who have neither interest in nor ability for scientific research, to write papers and submit them to journals. The referees, who are themselves research scientists, are thus flooded with mediocre papers, about which they are required to produce reports. Since they have more interesting work to do, the reports are often hasty and superficial. Reasonable-looking papers are accepted, obviously bad papers are rejected, and good papers that are a bit original and out of the norm tend to be rejected too. This is a well-known problem, and nobody really knows what to do about it. Fortunately, there are many scientific journals, and a really good paper will eventually get published somewhere.

6. J. C. Roux, A. Rossi, S. Bachelart, and C. Vidal, "Representation of a strange attractor from an experimental study of chemical turbulence," *Phys. Letters* 77 A (1980): 391–93.

7. D. Ruelle, "Large volume limit of the distribution of characteristic exponents in turbulence," *Commun. Math. Phys.* 87 (1982): 287–302.

The problem of cheating in scientific research is a delicate one. The traditional view was that cheating is exceptional, and that research scientists have very high ethical standards (with a few exceptions). That traditional view is now challenged, and cheating is openly discussed as an important factor in the quality of science. Let me briefly present the two main areas in which cheating occurs: *priorities* and *faking of data*.

The priority question is "Who did it first?" A fine example of a priority dispute is that between Newton and Leibniz on the invention of the differential calculus. If you are a conscientious scientist, you will acknowledge the sources of all the ideas that you use (supposing you remember). If you are unscrupulous, you will try to present as your own some results obtained by others. For example, if you find a good idea in a paper that you referee, you will try to stop that paper, and rush to publish the idea under your own name (or have one of your students publish it).

Much worse is the faking of data. Fraud of this kind has unfortunately been shown to occur on a large scale in biomedical research in the United States (for example, inventing clinical records for patients who do not exist). One reason for such fraud is that many people write papers for career reasons and have little interest in scientific truth. And there is the ever-present need to get results if funding is to be obtained.

I myself have worked in some areas in which I could freely discuss ideas with colleagues, and other areas in which it was unwise, because of the risk that the ideas would be stolen. The former is much more pleasant, *and allows faster scientific progress.*

Mathematics is relatively free from cheating, because it is a vast field with relatively few people working on any given problem. Faking of data does not occur, and the stealing of ideas is difficult because the ideas are complex. However, there are some priority disputes (remember Newton and Leibniz), there are some dubious characters around, and there is no guarantee that the present relatively satisfactory situation will last forever.

CHAPTER 12. CHAOS: CONSEQUENCES

1. M. Berry, "Regular and irregular motion," in *Topics in Nonlinear Dynamics: A Tribute to Sir Edward Bullard*, ed. S. Jorna (New York: American Institute of Physics, 1978), pp. 16–120. M. Berry's calculation (pp. 95–96) is based on earlier ideas of E. Borel and B. V. Chirikov. What is the gravitational effect of a distant body on the collision of two elastic balls? If the two balls are initially at different distances from the body, they will be deflected differently and the geometry of the collision will be slightly different depending on whether the body is present or not. Following a given ball, we see that the difference will be exponentially amplified in subsequent collisions. (The amplification is not by 2, as in our simplified discussion in Chapter 7, but by something like l/r, where l is the distance traveled by a particle and r is its radius.) After n collisions the angle between the original and the modified trajectories becomes of order 1, and the two trajectories no longer have anything to do with each other.

If the distant body is an electron at 10^{10} light-years and the elastic balls are oxygen molecules (at normal temperature and pressure), then $n = 56$. If the distant body is a human at 1 m from a billiard table and the elastic balls are billiard balls, then $n = 9$. This is of course according to classical mechanics. Quantum effects already make it impossible to aim properly, just once, an oxygen molecule at another oxygen molecule ($n = 0$). For billiard balls, quantum effects allow $n = 15$. (It would thus have been more reasonable to invoke quantum mechanics than classical mechanics for our argument, but it really doesn't make any difference for what follows.)

2. M. Berry's calculation referred to in the previous note shows that a tiny initial deflection drastically changes the structure of collisions between air molecules in a very short time. The microscopic structure of the

air, and the fluctuations occurring in it, have then become quite different. These so-called *thermal fluctuations* affect the density, velocity, and so on . . . of little volume elements of air (in which the number of molecules is not very large). We can estimate the time it takes for thermal fluctuations in a turbulent fluid to be amplified by sensitive dependence on initial condition to a macroscopic scale (say 1 cm). The calculation uses Kolmogorov's theory of turbulence. That theory (for dimensional reasons) gives an essentially unique value for the rate of growth of perturbations (the characteristic time for growth is proportional to the *turnover time* of the eddies of the macroscopic size selected). From microscopic fluctuations to macroscopic changes in turbulence takes about a minute (see D. Ruelle, "Microscopic fluctuations and turbulence," *Phys. Letters* 72 A [1979]: 81–82). Going from small scales to large scales of turbulence takes a time proportional to the turnover time of the largest eddies considered (again using Kolmogorov's theory, and dimensional arguments). We estimate that it takes a few hours or a day to reach the scale of kilometers. We now pass to the level of the circulation of the atmosphere over the entire planet, where the time it takes to amplify a small change to a globally different situation is estimated to be 1 or 2 weeks by meteorologists. (For a discussion of the meteorological problems involved see M. Ghil, R. Benzi, and G. Parisi, eds., *Turbulence and Predictability in Geophysical Fluid Dynamics and Climate Dynamics* [Bologna: Soc. Ital. Fis. (and Amsterdam: North Holland), 1985].)

The estimates just presented are rather insensitive to details (because the times estimated are logarithms, or based on dimensional arguments, and because the largest times come from the largest scales). Therefore, while one may question for instance the use of the Kolmogorov theory of turbulence, another theory is unlikely to give very different results.

3. See J. Wisdom, "Chaotic behavior in the solar system," *Proc. Royal Soc. London* 413 A (1987): 109–29. Each asteroid revolves around the sun on an ellipse, but the shape of the ellipse changes slowly, owing to the attraction of the heavy planet Jupiter. These changes of shape are important for certain *resonant* distances to the sun, or more precisely for certain values of the semimajor axis of the ellipse. (The semimajor axis is related to the period of revolution by Kepler's third law, and when the period of revolution of the asteroid is in resonance with the period of revolution of Jupiter, this planet has a strong perturbing effect; a resonance is said to occur when the two periods have a ratio p/q, where p and q are small integers.) Computer studies show that in the resonant case there is a chaotic variation with time of the shape of the asteroid's orbit

(i.e., of the ratio of small and large axes of the ellipse). When these variations are such that the asteroids can cross the orbit of the planet Mars, the asteroids disappear by collision, and a gap is formed in the belt. The calculations thus justify the observed fact that some resonances correspond to gaps, and others do not.

4. Early attempts to use quantitative methods in biology and the soft sciences have been excessively optimistic. In particular, it was thought that the dimension of many naturally occurring attractors could be obtained by a method called the *Grassberger-Procaccia algorithm* (P. Grassberger and I. Procaccia, "Measuring the strangeness of strange attractors," *Physica D* 9 [1983]: 189–208). The method works well on good long-time series, but gives misleading results for short series (D. Ruelle, "Deterministic chaos: the science and the fiction," *Proc. Royal Soc. London* 427 A [1990]: 241–48). For another idea, which appears very promising, see G. Sugihara and R. M. May, "Nonlinear forecasting as a way of distinguishing chaos from measurement error in time series," *Nature* 344 (1990): 734–41.

Chapter 13. Economics

1. For a collection of articles on economy and chaos, see P. W. Anderson, K. J. Arrow, and D. Pines, eds., *The Economy as an Evolving Complex System* (Redwood City, Calif.: Addison-Wesley, 1988). At the meeting at which this book originated, both economists and physicists were present, and it is of interest that the economists were, as a rule, rather more prudent in their claims than the physicists. See also note 4 of Chapter 12.

Chapter 14. Historical Evolutions

1. W. B. Arthur, "Self-reinforcing mechanisms in economics," pp. 9–31 in *The Economy as an Evolving Complex System* (the book referred to in note 1 of Chapter 13).

Chapter 15. Quanta: Conceptual Framework

1. R. P. Feynman, *QED* (Princeton: Princeton University Press, 1985). Feynman's presentation of quantum mechanics is somewhat different from the more traditional presentation that we shall discuss, but in principle equivalent to it.

2. Remember that a *complex* number is a mathematical object of the form $z = x + iy$, where x and y are *real* numbers (like 1.5, or π, or -3), and the square $i^2 = i \times i$ of i is -1. One can compute with complex numbers much the same way one does with real numbers. The *complex conjugate* of z is $\bar{z} = x - iy$; it is easy to check that $z\bar{z} = x^2 + y^2$, and one writes $|z|$ = positive square root of $z\bar{z}$. Complex numbers are a bit less intuitive than real ones, but have some technical advantages. For instance, complex numbers always have (complex) square roots.

3. This note and the next two notes give a very quick overview of quantum mechanics.

The Schrödinger equation.

Remember that Newton's equation was (note 1 of Chapter 5)

$$m_j \frac{d^2}{dt^2} x_j = F_j \qquad \text{for } j = 1, \ldots, N.$$

We shall assume that there is a function V of $x_1, \ldots, x_N$ (called the *potential* function) such that

$$F_j = - \operatorname{grad}_{(j)} V,$$

where $\operatorname{grad}_{(j)}$ is the vector of derivatives with respect to the components of the position x_j of the jth particle. (For the case of gravitational interactions we have

$$V(x_1, \ldots, x_N) = -\gamma \sum_{j<k} \frac{m_j m_k}{|x_k - x_j|}.)$$

In quantum mechanics there is an amplitude $\psi(x_1, \ldots, x_N; t)$ for finding our N particles at positions $x_1, \ldots, x_N$ (at time t), and the amplitudes ψ form what is called the *wave function*. The time evolution of ψ is obtained by solving the Schrödinger equation

$$\frac{ih}{2\pi} \frac{\partial}{\partial t} \psi = -\frac{h^2}{8\pi^2 m} \sum_j \Delta_{(j)} \psi + V\psi,$$

where i is the square root of -1, h is some constant (Planck's constant), and $\Delta_{(j)}$ is the *Laplacian* with respect to x_j, i.e., $\Delta_{(j)} \psi$ is the sum of the second partial derivatives of ψ with respect to the components of x_j.

It is assumed that the $3N$-dimensional integral

$$\int |\psi(x_1, \ldots, x_N; t)|^2 \, dx_1 \ldots dx_N = 1$$

for some value of t, and this property then is true for all t.

4. A linear operator A acting on a function ϕ of $x_1, \ldots, x_N$ produces a new function $A\phi$ of these variables, in such a way that $A(c_1\phi_1 + c_2\phi_2) = c_1 A\phi_1 + c_2 A\phi_2$, where c_1 and c_2 are constants and ϕ_1, ϕ_2 two functions. Write now

$$(\phi_1, \phi_2) = \int \overline{\phi}_1 (x_1, \ldots, x_N)\, \phi_2 (x_1, \ldots, x_N)\, dx_1 \ldots dx_N,$$

where $\overline{\phi}_1$ is the complex conjugate of ϕ_1. [We always use functions ϕ such that (ϕ, ϕ) is finite.] If the linear operator A satisfies

$$(\phi_1, A\phi_2) = (A\phi_1, \phi_2),$$

then A is said to be *self-adjoint*, and such operators are suitable to correspond to physical observables.

For instance, the observable A corresponding to the first component x_{j1} of the position of the jth particle satisfies

$$(A\phi)(x_1, \ldots, x_N) = x_{j1}\, \phi(x_1, \ldots, x_N)$$

(the product of x_{j1} and ϕ). The observable v_j corresponding to the velocity of the jth particle is such that

$$(v_j\phi)(x_1, \ldots, x_N) = \frac{1}{m_j} \cdot \frac{h}{2\pi i} \operatorname{grad}_{(j)} \phi(x_1, \ldots, x_N).$$

Finally, the mean value of A at time t is defined to be

$$\begin{aligned} \langle A \rangle &= (\psi, A\psi) \\ &= \int \overline{\psi}(x_1, \ldots, x_N; \mathrm{t})\, (A\psi)(x_1, \ldots, x_N; t)\, dx_1 \ldots dx_N, \end{aligned}$$

where ψ is the wave function of our system. (This is the definition of the mean value for the *vector state* defined by the wave function ψ. There are more general mean values defined by *density matrices*, and corresponding more closely to the probability distributions of classical probability theory.)

5. If the self-adjoint operator A satisfies $A^2 = A$, it is called a *projection*, and such operators are suitable to correspond to *simple events*. Given two linear operators A and B, their product AB is the linear operator such that $(\mathrm{AB})\phi = A(B\phi)$ for all functions ϕ. In particular, if $AB = BA$ we say that A and B are *commuting* operators. The product AB of two commuting projections is again a projection, and is suitable to represent the event "A and B" if A and B represent the events "A" and "B." If AB

$\neq BA$, there is no natural definition of a projection corresponding to the problematic event "A and B."

A *complex event* in which several detectors click or do not click corresponds to a self-adjoint operator that need not be a projection (but it is positive, i.e., it is the square of a self-adjoint operator). Here again one can define "A and B" when A and B commute.

6. It is fair to say that Bell's ideas do not quite agree with those sketched in the present chapter; see J. S. Bell, *Speakable and Unspeakable in Quantum Mechanics* (Cambridge: Cambridge University Press, 1987). (This is a collection of reprinted papers by Bell, and it has been very well received by the physics community.)

7. See footnote 8 on p. 76 of *QED* (the book referred to in note 1 of Chapter 15). The collapse of wave packets is one of the attempts to put more in the mathematical formalism of quantum mechanics than is strictly needed to account for experimental evidence. There is nothing wrong with such attempts *provided they remain compatible with experimental evidence*. Other kinds of enlargement of the mathematical formalism of quantum theory have been proposed by David Bohm (see Bell's book in note 6 of Chapter 15) and in R. B. Griffiths, "Consistent histories and the interpretation of quantum mechanics," *J. Statist. Phys.* 36 (1984): 219–72.

Chapter 16. Quanta: Counting States

1. A serious technical discussion would add a grain of salt to our analysis: it is impossible to have the position strictly limited to an interval $[0,L]$ and the velocity to an interval $[-v_{max},v_{max}]$ if L and v_{max} are finite (technically, because the Fourier transform of a wave function ψ with compact support cannot have compact support if $\psi \neq 0$). But one can arrange that the probabilities for being outside $[0,L]$ or $[-v_{max},v_{max}]$ are very small. Physicists know that discussions involving little rectangles, as we have done, are not quite correct. But they are convenient and often give the right answer. What one has to keep in mind is that quantum mechanics is not just a statistical theory based on the Heisenberg uncertainty relations, although this way of looking at things often gives the correct answers to simple questions.

2. For N particles in a volume V with total kinetic energy at most E, we use the formula

$$\text{number of states} = \frac{1}{N!}\, S_{3N} \left(\frac{1}{h^3} V (2mE)^{3/2}\right)^N.$$

This is the volume in phase space, in units of h^{3N}, divided by the permutation number $N!$ to take into account the indistinguishability of particles [h = 6.6 E(-34) Joules × sec is Planck's constant, S_{3N} is the volume of the $3N$-dimensional sphere of radius 1, and m is the mass of a particle; here m = 7E(-27) kg, V = E(-3)m^3, N = 2.7E22]. We take $E = 3NkT/2$ [k = 1.4E(-23) Joules/(degree Kelvin) is Boltzmann's constant, and T is the absolute temperature, here 300 degrees Kelvin]. Therefore

$$\text{number of states} \approx \frac{1}{h^{3N}} \left(\frac{V}{N}\right)^N (2\pi mkT)^{3N/2} e^{5N/2}$$
$$\approx 1\text{E}500000000000000000000000.$$

We have ignored the technical problems of quantum statistics and of spin; these are not essential for the present discussion.

Chapter 17. Entropy

1. The *first law of thermodynamics* asserts that energy is conserved in all processes. (This is true provided one takes into account all forms of energy, including heat.)

Chapter 18. Irreversibility

1. *Ergodicity.*

Consider N atoms of helium in a one-liter container as forming a classical mechanical system (the helium atoms collide with the walls of the container, and we also allow them to interact with each other). For each atom let x_i be its position and mv_i the product of its mass and velocity (= momentum). The collection X of x_i, mv_i is a point in the phase space M of our system. After a time t, X is replaced by a new point f^tX, and f^tX has the same total energy as X. Call *energy shell* the set M_E of X's with the same energy E. The volume in phase space (product over i of the dx_i and mdv_i) induces in a natural way a volume on the energy shell. If A is a subset of M_E, and vol A its volume, then

$$\text{vol}(f^tA) = \text{vol}\, A,$$

i.e., the volume is preserved by time evolution. It would require a little care to formulate all of this precisely (for instance, A must be assumed to be measurable), but everything so far is rather straightforward. Here now is something new. We say that the time evolution on the energy shell M_E

is *ergodic* if an invariant subset J of M_E (i.e., $f^tJ = J$ for all t) cannot be such that $0 < \text{vol } J < \text{vol } M_E$ (i.e., J must have zero or full volume).

Suppose that the time evolution f^t is ergodic. Then for almost every initial condition X, and for every subset A of M_E, the fraction of time spent by f^tX in A is $\text{vol}A/\text{vol}M_E$. [More precisely, if $l(X,A,T)$ is the length of time spent by f^tX in A with $0 < t < T$, then $\lim l(X,A,T)/T = \text{vol}A/\text{vol}M_E$ when $T \to \infty$; this is a form of the *ergodic theorem*.] For ergodic time evolutions, then, the time averages are simply related to volumes in the energy shell, and this is why ergodicity is so important. Unfortunately, it is very difficult to prove that a mechanical system is ergodic. This has been achieved for Sinai's billiards of Chapter 7, but for very few other systems of interest. For our system of helium atoms it remains a case of the "ergodic hypothesis."

2. For an ergodic time evolution we have long return times, and this gives an explanation of irreversibility. But we can have irreversibility without ergodicity provided only that the return times are long. Some weakening of the ergodicity requirement is thus possible, and may be necessary in some physical theories. I mentioned in Chapter 17 that sensitive dependence on initial condition is useful for understanding irreversibility. How does that work? Actually, sensitive dependence on initial condition is not necessary for ergodicity, but it helps, and is for instance the first step in the proof of ergodicity for Sinai's billiards.

For a time evolution that is not ergodic, a little bit of external noise will push the system from one "ergodic component" to another one provided that the energy shell is a connected set. This effect of small perturbations (like the gravitational effect of an electron at the limit of the known universe) is effective when there is sensitive dependence on initial condition, and will have the result that even a nonergodic system will look ergodic.

All this being said, it must be recognized that some mechanical systems refuse to behave ergodically. In fact, the KAM theory (due to A. N. Kolmogorov, V. A. Arnold, and J. Moser) gives important examples of violation of ergodicity. (For a general discussion of the KAM theory, see J. Moser, *Stable and Unstable Motions in Dynamical Systems*, Annals of Mathematics Studies no. 77 [Princeton: Princeton University Press, 1973].) Also, the numerical treatment of certain systems shows nonergodic behavior.

3. I. Prigogine, *From Being to Becoming* (San Francisco: Freeman, 1980). Incidentally, an important question is how and why it is that our universe started with so little entropy. A discussion of this point would

involve the big bang theory of the origin of the universe and would take us too far afield.

4. The invariance of the laws of physics under time reversal is in question only for the *weak interactions* of elementary particles. For those interactions, the time reversal operation T is not an exact symmetry, but another time reversing operation, TCP, is believed to be. Actually, most physicists think that these facts have little relevance to the irreversibility observed at the macroscopic level.

Chapter 19. Equilibrium Statistical Mechanics

1. W. Fucks and J. Lauter, *Exaktwissenschaftliche Musikanalyse*, Forschungsberichte des Landes Nordrhein-Westfalen no. 1519 (Köln-Opladen: Westdeutscher Verlag, 1965). I am indebted to Karine Chemla for this reference.

2. One aspect of this hard work is what has come to be known as the theory of *large deviations*. See D. Ruelle, "Correlation functionals," *J. Math. Phys.* 6 (1965): 201–20; O. Lanford, "Entropy and equilibrium states in classical statistical mechanics," in *Statistical Mechanics and Mathematical Problems*, Lecture Notes in Physics no. 20 (Berlin: Springer, 1973), pp. 1–113; R. S. Ellis, *Entropy, Large Deviations and Statistical Mechanics*, Grundlehren der Math. Wiss. no. 271 (New York: Springer, 1985).

3. The maximum of $S_I(E_I) + S_{II}(E_{II})$ subject to $E_I + E_{II} = E$ is the maximum over E_I of $S_I(E_I) + S_{II}(E - E_I)$, and occurs when the derivative of this quantity with respect to E_I vanishes. This gives $S'_I(E_I) - S'_{II}(E - E_I) = 0$, i.e., $T_I = T_{II}$.

Chapter 20. Boiling Water and the Gates of Hell

1. If you take molten glass instead of water, and allow it to cool, it will become more and more viscous, and in the end it will become quite rigid and solid cold glass. But physicists will tell you that glass is not a regular solid: its microscopic structure is not in equilibrium and will change if you wait long enough. It will not, however, change noticeably during your lifetime. What this means is that glasses are outside the piece of physical reality that is well described by equilibrium statistical mechanics.

2. See in particular D. Ruelle, *Statistical Mechanics: Rigorous Results* (New York: Benjamin, 1969); Ya. G. Sinai, *Theory of Phase Transitions: Rigorous Results* (Oxford: Pergamon, 1982).

3. See, for instance, D. J. Amit, *Field Theory, the Renormalization Group, and Critical Phenomena*, 2d ed. (Singapore: World Scientific, 1984), and references quoted therein.

4. A typical process in vacuum fluctuations is when an electron and a positron are created simultaneously, and then very quickly afterwards annihilate each other and disappear. (An electron cannot spring out of the vacuum alone, or disappear, because of charge conservation.) Processes like this one are studied in *quantum electrodynamics* (QED), and Feynman's book gives an accessible introduction to this fascinating area of physics (see note 1 of Chapter 15).

5. An interesting book on black holes is K. S. Thorne, R. H. Price, and D. A. Macdonald's, *Black Holes: The Membrane Paradigm* (New Haven: Yale University Press, 1986). This is a technical book, full of complicated formulas, but the display of the complicated formulas is useful to physicists who want to get an idea of what the dirty technical details of the theory are like, even if they do not want to become experts themselves. Very much more accessible is of course Hawking's own popular account: S. W. Hawking, *A Brief History of Time* (London: Bantam, 1988).

Chapter 21. Information

1. Since the AIDS virus is an RNA virus, the four letters are not A, T, G, C originally, but a copy into this alphabet is made by reverse transcriptase.

2. For a family of messages with probabilities $p_1, p_2, \ldots$ the mean information content of one message is

$$\text{mean information content} = -\sum_i p_i \log p_i.$$

If there are N messages with probability $1/N$ each, the mean information is thus $\log N$. In many cases, the *Breiman-McMillan theorem* reduces the study of messages with different probabilities to the study of equiprobable messages. For a good technical discussion of information theory, including the Breiman-McMillan theorem, see P. Billingsley, *Ergodic Theory and Information* (New York: John Wiley, 1965).

3. C. Shannon, "A mathematical theory of communication," *Bell System Tech. J.* 27 (1948): 379–423, 623–56.

4. To study the information content of a melody, one would like to have the statistics corresponding to groups of 2,3,4, . . . consecutive notes. But intervals between 2 consecutive notes provide a convenient overestimate of the information.

5. See the reference in note 1 of Chapter 19. Of course, one has to compare musical pieces of the same length, or divide the information by the length of the piece.

6. In a specific discussion, the family of allowed messages should be indicated, for instance rectangular paintings with uniform color. (This class contains little information, because one can choose only the dimensions of the rectangle, and a particular color, and the number of choices that can be distinguished is not extremely large.) It may be difficult to specify explicitly the allowed family of messages in a given art form (like "abstract painting"), but we usually have some feeling for how little or how much freedom is available, in writing sonnets or novels for instance.

Chapter 22. Complexity, Algorithmic

1. See M. R. Garey and D. S. Johnson, *Computers and Intractability* (New York: Freeman, 1979). This is the standard reference for algorithmic complexity and contains in particular a discussion of Turing machines.

2. An efficient algorithm for linear programming has been invented by L. G. Khachiyan, and a more practical one by N. Karmarkar. See note 1 of Chapter 6 for the formulation of finite zero-sum two-person game problems as linear programming problems.

3. NP stands for *Nondeterministic Polynomial*. This is because (as discussed below) a positive answer can be verified in polynomial time if a correct guess has been made (nondeterministically). The NP complete problems are all equally difficult: if you can solve one you can solve all, hence the qualification *complete*.

4. For spin glasses and disordered systems see M. Mézard, G. Parisi, and M. A. Virasoro, *Spin Glass Theory and Beyond* (Singapore: World Scientific, 1987). The spin glass problem, as we have defined it, is not discussed in Garey and Johnson (note 1, Chapter 22), but is close to SMC ("Simple Max Cut"), which is known to be NP complete.

5. The tree structure of natural evolution is analogous to the tree structure of *valleys* in the Parisi solution of the spin glass model (for this see

Spin Glass Theory and Beyond in the preceding note). This analogy seems to be maintained at the quantitative level (see H. Epstein and D. Ruelle, "Test of a probabilistic model of evolutionary success," *Physics Reports* 184 [1989]: 289–92).

Chapter 23. Complexity and Gödel's Theorem

1. I heard the story from R. V. Kadison.

2. The following book (in French) is very helpful in getting oriented in Freud's work: J. Laplanche and J.-B. Pontalis, *Vocabulaire de la psychanalyse* (Paris: PUF, 1967).

3. What does it mean that an assertion is true if it cannot be proved or disproved from the axioms? To see this it is necessary to understand the nature of the game called *metamathematics*, played by mathematical logicians. The mathematicians have various theories *A*, *B*, . . . , each based on a system of axioms that is believed to be noncontradictory. For instance, *A* might be an axiomatic presentation of the arithmetic of integers, and *B* of set theory. (Gödel showed that one cannot prove the noncontradiction of the sort of axiom systems used by mathematicians. Some faith is therefore needed here. But most mathematicians are quite convinced that no contradiction will ever arise from the axioms of arithmetics or set theory which they are using.) The axioms, theorems, and rules of inference of theory *A* may now be viewed as mathematical objects to which theory *B* can be applied. One is thus looking at theory *A from the outside*, so to say, and it is possible in this manner to prove things about it that are inaccessible *from the inside*. This is the metamathematical game, and it is tricky. But if you believe in the noncontradicion of *A* (and of *B*), consequences like Gödel's incompleteness theorem are inescapable.

4. R. J. Solomonoff, "A formal theory of inductive inference," *Inform. and Control* 7 (1964): 1–22, 224–54; A. N. Kolmogorov, "Three approaches to the definition of the concept 'quantity of information,' " *Probl. Peredachi Inform.* 1 (1965): 3–11; G. J. Chaitin, "On the length of programs for computing finite binary sequences," *J. ACM* 13 (1966): 547–69. See also G. J. Chaitin, *Algorithmic Information Theory* (Cambridge: Cambridge University Press, 1987); G. J. Chaitin, *Information, Randomness, and Incompleteness* (Singapore: World Scientific, 1987).

5. See Theorem 2 in the Appendix of G. J. Chaitin, "Information-theoretic computational complexity," *IEEE Trans. Inform. Theory* IT-20

(1974): 10–15. This paper is reprinted (pp. 23–32) in *Information, Randomness, and Incompleteness* (see preceding note).

6. See M. Davis, Y. Matijasevič, and J. Robinson, "Hilbert's tenth problem. Diophantine equations: Positive aspects of a negative solution," in *Mathematical Developments Arising from Hilbert Problems*, Proc. Symp. Pure Math. 27 (Providence, R.I.: American Mathematical Society, 1976), pp. 323–78.

7. See the book *Algorithmic Information Theory* referred to in note 4 of Chapter 23. Chaitin's sequence actually becomes random only after a finite number of terms.

8. One suggestion by Pierre Cartier is that the axioms of set theory are actually inconsistent, but that a proof of contradiction would be so long that it could not be performed in our physical universe! More conservatively, we may expect that further developments of mathematical logic will be compatible with what is currently accepted, but will shed new light on the foundations of mathematics.

Chapter 24. The True Meaning of Sex

1. We may assume that the offspring of a message is proportional to exp[E(message)], and allow mutations from a message to closely similar messages. The basic defect of this model, or metaphor, of life is that it does not capture the dynamical aspects of the relations of a message with messages of the same kind and of different kinds (i.e., population dynamics is not taken into account).

2. For mathematical simplicity, we think here of point mutations (although other types of mutations have great evolutionary importance). Point mutations correspond to a random walk in the random environment provided by the function E. The assumption that the offspring of a message is proportional to exp[E(message)] means that large values of E are favored. Random walks in a random environment are known to proceed very slowly, because to go from one mountain to another one, it is first necessary to climb down, and this is a very unlikely process (see Ya. G. Sinai, "Limit behavior of one-dimensional random walks in random environments," *Teor. Verojatn. i ee Primen.* 27 [1982]: 247–58; English translation: *Theor. Probab. Appl.* 27 [1982]: 247–58; E. Marinari, G. Parisi, D. Ruelle, and P. Windey, "On the interpretation of $1/f$ noise," *Commun. Math. Phys.* 89 [1983]: 1–12; R. Durrett, "Multidimensional random walks in random environments with subclassical limiting behavior," *Commun. Math. Phys.* 104 [1986]: 87–102. The random walk

therefore tends to be trapped on top of small mountains. This could be avoided by increasing the mutation rate, but such an increase is severely restricted by the necessity of retaining meaningful genetic messages. In fact, as one goes from simple organisms with short genetic messages to complex organisms with long genetic messages, one finds more and more accurate replication mechanisms, which reduce mutations to lower and lower levels. That this should be so is understandable from an information-theoretic viewpoint. (See M. Eigen and P. Schuster, *The Hypercycle: A Principle of Natural Self-Organization* [Berlin: Springer, 1979].) Altogether we see why evolution uses many other tricks than just point mutations (increase in or deletion of genetic material, sex, and symbiosis are important for evolution).

3. Sex is not universal among living organisms, but it is very common. Some bacteria have genetic recombination, and therefore sex. This does not mean that there are always two different genders (this is a less essential innovation, however important it is to us).

4. It is usually accepted that sex helps evolution, but there are dissenting voices. See L. Margulis and D. Sagan, *Origins of Sex* (New Haven: Yale University Press, 1986).

5. R. Dawkins, *The Selfish Gene* (Oxford: Oxford University Press, 1976).

6. The earth was formed 4.5E9 years ago, and rocks 3.5E9 years old show evidence of life. By geological standards, it seems that life formed essentially as soon as the environmental conditions permitted it. Let us remark in passing that the function E(message) at the time was rather different from what it has now become.

Chapter 25. Intelligence

1. D. Marr, *Vision* (New York: Freeman, 1982).

2. The processes in which Freud is interested are *mind* processes.

3. It is, of course, an idealization to see our world as three-dimensional, and containing objects limited by surfaces. Scientists use many other idealizations as well, but this particular idealization was encouraged by evolution and has been hard-wired into our brains. It is an idealization that has served us well, both for survival and for the development of geometry and other sciences.

4. E. Wigner, "The unreasonable effectiveness of mathematics in the natural sciences," *Commun. Pure Appl. Math.* 13 (1960): 1–14.

Chapter 26. Epilogue: Science

1. A curious and interesting essay should be mentioned here: R. Penrose, *The Emperor's New Mind* (New York: Oxford University Press, 1989). This is a brilliant exposition of modern scientific ideas. At the same time it is an elaborate plea, suggesting that the laws of physics should be altered to accommodate consciousness, and the introspective view that our minds do not function as computers. Clearly, the laws of physics will have to be changed to accommodate quantum gravity, but I very much doubt that this will agree with Penrose's ideas. When dealing with consciousness and introspective certainties, we should always remember how clever and how forceful our minds are at self-deception. This is one lesson of psychoanalysis that cannot be lightly dismissed.